John Hunter, Everard Home, Thomas Bradford

A Treatise on the Blood, Inflammation and Gun-Shot Wounds

Volume 2

John Hunter, Everard Home, Thomas Bradford

A Treatise on the Blood, Inflammation and Gun-Shot Wounds
Volume 2

ISBN/EAN: 9783337392666

Printed in Europe, USA, Canada, Australia, Japan

Cover: Foto ©berggeist007 / pixelio.de

More available books at **www.hansebooks.com**

ON

THE BLOOD,

INFLAMMATION,

AND

GUN-SHOT WOUNDS,

BY THE LATE

JOHN HUNTER.

TO WHICH IS PREFIXED

A SHORT ACCOUNT OF THE AUTHOR's LIFE,

BY HIS BROTHER-IN-LAW,

EVERARD HOME.

IN TWO VOLUMES, FROM THE LONDON QUARTO.

VOL. II.

PHILADELPHIA :

PUBLISHED BY THOMAS BRADFORD, PRINTER,
BOOK-SELLER & STATIONER,
No. 8,
South Front-Street,
1796.

TABLE OF CONTENTS.

PART II.

CHAPTER III.

Table of contents.

CHAPTER VII.

CHAPTER VIII.

CHAPTER IX.

PART III.
CHAPTER I.

PART IV.

CHAPTER I.

CHAPTER II.

CHAPTER III.

THE ADHESIVE INFLAMMATION.

I SHALL begin by treating of the nature and effects of what I have called the adhesive inflammation, as well as giving a proper idea of it. I shall also open the way to a clear understanding of the many phænomena which attend the suppurative inflammation ; but as inflammation does not produce one effect only, but several, and as most of them take place about the same time, it is difficult to determine in the mind, which to describe first.

Inflammation in most cases appears to begin at a point ; for at the very first commencement, all the local symptoms are within a very small compass, and they afterwards spread according to the violence of the cause ; the disposition in the parts for inflammation, and the nature of the surrounding parts themselves ; which susceptibility in the surrounding parts may be either constitutional or local.

This is so much the case, that inflammation shall come on at once in a fixed point, giving great pain, and which shall be soon followed by tumefaction.

This is also the case with those inflammations which arise from accident, for all accidents are confined to fixed and determined limits, but the inflammation which follows is not ; it spreads over a large extent, yet the inflammation is always the greater, the nearer it is to the first fixed point ; and gradually becomes less and less in the surrounding parts, till at last it is insensibly lost in them.

This spreading of the inflammation is owing to continued sympathy, the surrounding parts sympathising with the point of irritation ; and in proportion to the health of the surrounding parts and constitution, this sympathy

4

is lefs ; for we find in many ftates of parts, and many conftitutions, that there is a difpofition to this fympathy, and in fuch, the inflammation fpreads in proportion.

I. ACTION OF THE VESSELS IN INFLAM-MATION.

THE act of inflammation would appear to be an in-creafed action of the veffels *, but whatever action it is, it takes place, moft probably in the fmaller veffels, for it may be confined almoft to a point where nothing but the fmalleft veffels exift. The larger veffels may be confider-ed as only the conveyors of the materials, for the fmaller to act upon and difpofe of according to the different in-tentions ; however, inflammation in a part, is not only an action of the fmaller veffels in the part itfelf, but in the larger veffels leading to it. This is proved by a whitlow taking place on the end of a finger ; for although the in-flammation itfelf fhall be confined to the end of a finger, and the inflammatory fenfation or throbbing be fituated in this part, yet we can feel by our hands, when we grafp the finger, a ftrong pulfation in the two arteries leading to the inflamed part, while no fuch pulfation can be felt in the other fingers ; and if the inflammation is very con-fiderable, the artery, as high as the wrift, will be fenfibly affected, which proves that the arterial fyftem is at that time dilating itfelf, and allowing a much larger quantity of blood to pafs than is ufual. This is probably by conti-nued fympathy.

Where the inflammation affects the conftitution, the veffels of the fyftem rather contract, and keep as it were ftationary, which ftationary contraction is more or lefs ac-cording to the ftate of the conftitution ; in ftrong healthy

* It may be here remarked, that the action of veffels is commonly fuppofed to be contraction, either by their elaftic or mufcular coats ; but I have fhewn that their elaftic power alfo dilated them ; and I have reafon to believe the mufcular power has a fimilar effect.

conftitutions whofe powers are equal to the neceffary actions, or in parts that affect the conftitution lefs, this contraction is lefs and lefs ftationary.

The very firft act of the veffels when the ftimulus which
excites inflammation is applied, is, I believe, exactly fimilar to a blufh. It is, I believe, fimply an increafe or
diftenfion beyond their natural fize. This effect we fee
takes place upon many occafions, gentle friction on the
fkin produces it ; gently ftimulating medicines have the
fame effect, a warm glow is the confequence, fimilar to
that of the cheek in a blufh : and if either of thefe be increafed or continued, real inflammation will be the confequence, as well as excoriation, fuppuration and ulceration. This effect we often fee, even where confiderable
mifchief has been done ; and I believe it is what always
terminates the boundaries of the true inflammation. A
mufket-ball fhall pafs a confiderable way under the fkin,
perhaps half way round the body, which fhall be difcovered and traced by a red band in the fkin, not in the leaft
hard, only a little tender to the touch ; and it fhall fubfide
without extending farther. This appearance I fhall term
a blufh ; for although this may be reckoned the firft act of
inflammation, yet I would not call it inflammation, having produced a lafting effect ; I fhould rather fay, that
inflammation fets out from this point, and that afterwards
a new action begins, which is probably firft a feparation
of the coagulating lymph, and the throwing it out of the
veffels.

The parts inflamed appear to become more vafcular ;
but how far they are really fo, I am not certain, for this
appearance does (at leaft in part) arife from the dilation of
the veffels, which allows the red part of the blood to go into veffels where only ferum and coagulating lymph could
pafs when they were in a natural ftate, and ftill the newly extravafated fubftances become vafcular ; the effect is
moft probably owing wholly to the above caufe.

This incipient enlargement of the veffels upon the firft
excitement of inflammation is fatisfactorily feen in the following manner. Make an incifion through the fkin on the
infide of the upper part of a dog's thigh, three inches
long ; by pulling the cut edges afunder, and obferving the
expofed furface, we fhall fee the blufh or afh-coloured
cellular membrane covering the different parts underneath,

with a few arteries paffing through it to the neighbouring parts; but in a little time we fhall fee thefe veffels increafing in fize, and alfo fmaller veffels going off from
them that were not before obfervable, as if newly formed
or forming; the number and fize fhall increafe till the
whole furface fhall become extremely vafcular, and at laft
the red blood fhall be thrown out in fmall dots on the expofed furface, probably, through the cut ends of the arteries that only carried the lymph before.   This furface
will become in time more opaque, and lefs ductile.   Parts
inflamed, when compared with fimilar parts not inflamed,
fhew a confiderable difference in the fize of the veffels, and
probably from this caufe bring an increafed number to
view.   I froze the ear of a rabbit and thawed it again;
this excited a confiderable inflammation, and increafed
heat, and a confiderable thickening of the part.   This
rabbit was killed when the ear was in the height of inflammation, and the head being injected, the two ears were
removed and dried.   The unimflamed ear dried clear and
tranfparent, the veffels were diftinctiy feen ramifying
through the fubftance; but the inflamed ear dried thicker
and more opaque, and its arteries were confiderably larger.

In inflammation of the eye, which is commonly of the
tunica conjunctiva, the progrefs of inflammation may, in
part, be accurately obferved, although not fo progreffively
as in a wound.   The contraft between the red veffel and
the white of the eye, under this coat is very confpicuous,
and although we do not fee the veffels enlarging in this
coat, yet we fee the progrefs they have made, the white
appears as if it was becoming more vafcular, and thefe veffels larger, till at laft the whole tunica conjunctiva fhall
appear as one mafs of blood, looking more like extravafated
blood then a congeries of veffels, although I believe it is
commonly the laft.

From thefe circumftances it muft appear, that a much
larger quantity of blood paffes through parts when inflamed than when in a natural ftate, which is according to
to the common rules of the animal œconomy; for, whenever a part has more to do than fimply to fupport itfelf,
the blood is there collected in larger quantity; this we
find to take place univerfally in thofe parts whofe powers
are called up to action by fome neceffary operation to be
performed, whether natural or difeafed.

As the veffels become larger, and the part becomes of the colour of the blood, it is to be fuppofed there is more blood in the part ; and as the true inflammatory colour is fcarlet, or that colour which the blood has when in the arteries, one would from hence conclude, either that the arteries were principally dilated, or at leaft, if the veins are equally diftended, that the blood undergoes no change in fuch inflammation in its paffage from the arteries into the veins, which I think is moft probably the cafe ; and this may arife from the quicknefs of its paffage through thofe veffels.

When inflammation takes place in parts that have a degree of tranfparency, that tranfparency is leffened. This is, probably, beft feen in membranes, fuch as thofe membranes which line cavities, or cover bodies in thofe cavities, fuch as the pia-mater, where, in a natural ftate, we may obferve the blood-veffels to be very diftinct. But when we fee the blood-veffels fuller than common, yet diftinct in fuch membranes, we are not to call that inflammation, although it may be the firft ftep, as we find to be the cafe in the firft action of the veffels in confequence of fuch irritation as will end in inflammation. As it may not, however, be the firft ftep, there muft be other attending circumftances to determine it to be the very firft action of the veffels in inflammation, for as that appearance may either belong to a brifknefs in the circulation in the part at the time, or the very firft ftep in inflammation, their caufes are to be difcriminated by fome other fymptom ; they are both a kind of blufh, or an exertion of the action of the veffels ; but when it is an effect of an inflammatory caufe, it is then only that the inflammation has not yet produced any change in the natural ftructure of the parts, but which it will foon do.‡ What the action is, or in what it differs from the common action of the veffels, is not eafily afcertained, fince we are more able to judge of the effects than the imme-·

‡ When this appearance is feen any where after death, it fhould not be called inflammation, even although we knew it was the firft action of inflammation ; for as we are then only looking out for the caufes of death, or the fymptoms prior to death, we are only to look out for fuch as can be a caufe, and not lay hold of thofe that cannot poffibly be a caufe, which thofe firft actions cannot be.

diate caufe.　　However, it is probably an action of the
the veſſels, which we can better obſerve h..: ov diſeaſed
action in the body, for we ca. (bſ.iv. ti.. !.te in which
the arteries are, with their ... neral effe.. s.; we te.i, alſo,
a different temperature re'.... uig he t, ye. the immediate
caufe may not be aſcertainable.

The veſſels, both arteries and veins, in the inflamed part
are enlarged, and the part beccmes viſibly more vaſcular,
from which we ſhould ſuſpect, that inſtead of an increaſed
contraction, there was rather what would appear an increaſed relaxation of their muſcular powers, being, as we
might ſuppoſe, left to the elaſticity entirely.　This would
be reducing them to a ſtate of paralyſis ſimply ; but the
power of muſcular contraction would ſeem to give way in
inflammation, for they certainly dilate more in inflammation than the extent of the elaſtic power would allow; and it
muſt alſo be ſuppoſed, that the elaſtic power of the artery
muſt be dilated in the ſame proportion.　The contents of
the circulation being thrown out upon ſuch occaſions,
would, from conſidering it in thoſe lights, rather confirm
us in that opinion ; and when we conſider the whole of
this as a neceſſary operation of nature ; we muſt ſuppoſe it
ſomething more than ſimply a common relaxation ; we muſt
ſuppoſe it an action in the parts to produce an increaſe of ſize to
anſwer particular purpoſes; and this I ſhould call the action
of dilatation, as we ſee the uterus increaſe in ſize in the time
of uterine-geſtation, as well as the os tincæ in the time
of labour, the conſequence of the preceding actions, and
neceſſary for the completion of thoſe which are to follow.

The force of the circulation would ſeem to have ſome
ſhare in this effect, but only as a ſecondary cauſe; for I could
conceive a part to inflame, or be in a ſtate of inflammation, although no blood were to paſs.　As a proof of this,
we may obſerve, that by leſſening either the action of the
heart, or the column of blood, inflammation is leſſened ;
and I may alſo obſerve, that we have an increaſed pain in
the inflamed part in the diaſtole of the artery, and a part
inflamed by being gently preſſed is made eaſier.　Thus a
perſon with an inflammation in the fingers will find relief
by gently preſſing it in the other hand.　Theſe are ſtrong
proofs that it is not a contractile action of the vaſcular
coat of the veſſel ; for in ſuch a ſenſible ſtate of veſſels if
they contracted by their muſcular power, the pain would
be in their ſyſtole ; for we find in all muſcles which are in

a ftate of great fenfibility, from whatever caufe, that they cannot act without giving great pain. Thus an inflamed bladder becomes extremely painful when expelling its contents, an inflamed inteftine in the fame manner; I fhould fay, therefore, that in inflammations the mufcular coats of the arteries do not contract.

Whatever purpofe this increafe of the fize of the veffels may anfwer, we muft fuppofe it allows a greater quantity of blood to pafs through the inflamed part than in the natural ftate, which fuppofition is fupported by many other obfervations.

The part inflamed, I have already obferved, becomes to appearance more vafcular than when in the natural ftate, and it is probable that it is really fo, both from new veffels being fet up in the inflamed part, as well as the new and adventitious uniting fubftance becoming vafcular. Befides, the veffels of the parts are enlarged, fo that the red blood paffes further than common, which increafes thofe appearances. But the brain appears to be an exception to thefe general rules ; for in all difeafes of the brain, where the effects were fuch as are commonly the confequence of inflammation, fuch as fuppuration from accidents, I never could find the above appearances ; the brain may, perhaps, go directly, into fuppuration, as fometimes the peritoneum does ; but its flownefs of going into fuppuration after the accident, would make us fuppofe, apriori, that there was fufficient time for adhefions to form.

―――――

II. OF THE COLOUR, SWELLING, AND PAIN OF INFLAMED PARTS.

THE colour of an inflamed part is vifibly changed from the natural, whatever it was, to red. This red is of various hues, according to the nature of the inflammation ; if healthy, it is a pale red ; if lefs healthy, the colour will be darker, more of a purple, and fo on till it fhall be a blueifh purple, which I took notice of in the fhort fketch of the peculiar inflammations ; but the parts inflamed will in every conftitution be more of the healthy red when the

Vol. II,   B

parts inflamed are near to the source of the circulation, than when far from it. This increase of red appears to arise from two causes; the first is a dilatation of the veffels, whereby a greater quantity of blood is allowed to pass into those veffels which only admitted ferum or lymph before.||

The fecond is owing probably to new veffels being fet up in the extravafated uniting coagulating lymph.

This colour is gradually loft in the furrounding part if the inflammation is of the healthy kind, but in many others it has a determined edge, as in the true eryfepelatous, and in fome fpecific difeafes, as in the fmall-pox, where its quick termination is a fign of health.

From the account I have given of the immediate effects of inflammation of the cellular membrane, in which I include the larger cavities, the volume of the part inflamed muft be increafed. This, when a common confequence of inflammation, is not circumfcribed, but rather diffufed, as the inflammation, however, begins in a circumfcribed part, which is at leaft the cafe with that arifing from violence ; the inflammation I juft now obferved is always the greateft neareft to that point, and is gradually loft in the furrounding parts, the fwelling of courfe is the greateft at, or neareft to this point, and it is alfo loft in the furrounding found parts. This takes place, more or lefs, according to the conftitution, or the fituation of the inflammation ; for if the conftitution be ftrong and healthy, the furrounding parts will fympathize lefs with the point of irritation, fo that inflammation and its confequences, viz. extravafation will be lefs diffufed.

There will be lefs of the ferum, and of courfe a purer coagulating lymph, fo that the fwelled parts will be firmer; but in fome fpecific difeafe or diffimilar part, as a gland, it has a more determined edge, the furrounding parts not fo readily taking on fpecific difeafed action as in other cafes. In this both the colour and fwelling correfpond very much fince they both depend on the fame principle.

The increafe of volume is owing to the extravafation of the coagulating lymph, and fome ferum ; in proportion to the inflammation, the degree of which depends on the

|| The tunica conjunctiva of the eye when inflamed, is a ftriking inftance of this ; but the vifible progrefs of inflammation I have already defcribed in the experiment on the dog.

caufes abovementioned, this effect is more or lefs, and therefore is greateft at the point of inflammation, becoming lefs and lefs as it extends into the furrounding parts, till it is infenfibly loft in them.

The extravafation of the ferum along with the coagulating lymph is, probably, not a feparation of itfelf, as in a dropfy, but a part of it being feparated from the lymph in the coagulation of that fluid, is fqueezed into the furrounding cellular membrane, where there is but little extravafation, and where the cells are not united by it. Thus the circumference of fuch fwelling is a little œdamatous ; but the whole of the ferum, if there be a depending part, will move thither, and diftend it confiderably, as in the foot in confequence of an inflammation in the leg. But in moft cafes there is a continued extravafation of ferum, long after the extravafation of the coagulating lymph is at at an end ; fo that depending parts will continue œdematous, while the inflammation is refolving, or while fuppuration, or even healing is going on.

The whole fwelling looks like a part of the body only a little changed, without any appearance of containing extraneous matter ; and indeed it is fimply formed by an extravafation of fluids without their having undergone any vifible or material change, except coagulation.

As few uncommon operations can go on in an animal body without affecting the fenfations, and as the firft principle of fenfation arifes from fome uncommon action, or alteration being made in the natural pofition or arrangement of the parts, we fhould naturally fuppofe that the fenfation would be in fome degree according to thofe effects, and the fenfibility of the parts. One can eafily form an idea of an alteration in the ftructure of parts giving fenfation which may even be carried to pain, but that the fimple action of parts fhould produce fenfations and even violent pain is but little known, or at leaft has been, I believe, but little attended to ; all thefe effects, I think, may juftly be included under the term fpafm‖; at leaft we are led by analogy to fuppofe that they belong to that clafs.

‖ How far a nerve from a part, or how far the materia vitæ of a part, can act fo as to convey fenfation I do not know ; but we all know that an involuntary action of a voluntary mufcle, or the fpontaneous action of an involuntary mufcle will produce it.

By fpafm I fhould underftand a contraction of a mufcle, without the leading and natural caufes.

Thus the contraction of a mufcle of the leg, called the cramp, gives confiderable pain, often violent, as alfo the tetanus, and when in a lefs degree, as in the twinkling of the eye-lid, it gives only fenfation, whereas if the mufcles were to act by the will, no fenfation would be produced.

We find that thofe fenfations are more or lefs acute, according to the quicknefs or flownefs of the progrefs of thefe caufes, from whence we are naturally led to affign two caufes which muft always attend one another; for when both do not take place at the fame time, the mind then remains infenfible to the alteration. This is its being produced in a given time, for the alteration in the pofition of the parts may be produced fo flowly, as not to keep pace with fenfation, which is the cafe with many indolent tumors, afcites, etc. on the other hand this alteration in the natural pofition of parts may be fo quick as to exceed fenfation, and therefore there is a certain medium, which produces the greateft pain.

The actions I have been defcribing being pretty quick in their effects we cannot fail to fee why the pain from the inflammation muft be confiderable; however the pain is not the fame in all the different ftages. In the adhefive ftate of the inflammation it is generally but very inconfiderable, efpecially if it is to go no further, and is perhaps more of a heavy than an acute pain; when it happens on the fkin it often begins with an itching; but as the inflammation is paffing from the adhefive to the fuppurative, the parts then undergo a greater change than before, and the pain grows more and more acute, till it become very confiderable. The nerves alfo acquire at that time a degree of fenfibility, which renders them much more fufceptible of impreffion than when they are in their natural ftate; thus an inflamed part is not only painful in itfelf, but it communicates impreffions to the mind independent of pain, which do not arife from a natural found part. This pain increafes every time the arteries are dilated, whence it would appear that the arteries do not contract by their mufcular power, in their fyftole, for if they did, we might expect a confiderable pain in that action which would be at the full of the pulfe. Whether this pain arifes from the diftention of the artery by the force of the heart, or whether it arifes from the action of diftention from the force of the artery itfelf, is not eafily determined. We know that difeafed mufcles give much pain in their contraction, perhaps more than they do when stretched*.

That the degree of inflammation which becomes the cause of adhesions gives but little pain, is proved from dissections of dead bodies; for we seldom or never find a body in dissection which has not adhesions in some of the larger cavities; and yet it may reasonably be supposed, that many of these persons never had any acute symptoms, or violent pain in those parts; indeed, we find many strong adhesions upon the opening of dead bodies, in parts which the friends of these persons never heard mentioned, during life, as the subject of a single complaint.

That adhesions can be produced from very slight inflammation, is proved in ruptures in consequence of wearing a truss; for we find the slight pressure of a truss exciting such action as to thicken parts, by which means the two sides of the sack are united, though there be hardly any sensation in the part; we also see, in cases where this inflammation arises from violence, it gives little or no pain. A man shall be shot through the cavity of the abdomen, and if none of the contained parts are materially hurt, the adhesive inflammation shall take place in all the internal parts contiguous to the wound made by the ball, and yet no great degree of pain shall be felt. This assertion is still proved by the little pain suffered after many bruises, where there is evident inflammation; and in simple fractures, the pain from the the inflammation is very trifling, whatever it may be from the laceration of the parts. But this will be according to the degree of inflammation, what stage it is in, and what parts are inflamed, as will be fully explained hereafter.

We find it a common principle in the animal machine, that every part increases in some degree according to the action required. Thus we find muscles increase in size when much exercised; vessels become larger in proportion to the necessity of supply, as for instance, in the gravid uterus; the external carotids in the stag, also, when his

---

* This is very evident in the bladder of urine when inflamed, for in the contraction of that viscus to expel the urine, there is more pain than in the dilatation; indeed, the distention is gradual, and when the urine is wholly evacuated, the irritation produced by the contraction still continues, which produces a continuance of the straining.

horns are growing, are much larger than at any other time; and I have obferved, that in inflammation, the veffels become larger, more blood paffes, and there appear to be more actions taking place; but the nerves do not feem to undergo any change. The nerves of the gravid uterus are the fame as when it is in the natural ftate; neither do the branches of the fifth and feventh pair of nerves in the ftag become larger; and in inflammation of the nerves, their blood-veffels are enlarged, and have coagulating lymph thrown into their inteftines, but the nerve itfelf is not increafed fo as to bring the part to the ftate of a natural part, fitted for acute fenfation, which fhews that the motions of the nerves have nothing to do with the œconomy of the part, they are only the meffengers of intelligence and order. It appears that only the actions of the materia vitæ in the inflamed parts is increafed, and this increafe of action in the inflamed part is continued along the nerve which is not inflamed, to the mind, fo that the impreffion on the fenforum is, probably, equal to the action of the inflamed materia vitæ.

The quantity of natural fenfibility is, I believe, proportioned to the quantity of nerves, under any given circumftance; but I apprehend, the difeafed fenfibility does not take place at all in this proportion, but in proportion to the difeafed action of the materia vitæ. Thus a tendon has very little fenfation when injured in a natural ftate; but let that tendon become inflamed, or otherwife difeafed, and the fenfation fhall be very acute.

It may not be improper to obferve, that many parts of the body in a natural ftate, give peculiar fenfations when impreffed; and when thofe parts are injured, they give, likewife, pain peculiar to themfelves; it is this latter effect, which I am to confider. I may alfo obferve, that the fame mode of impreffion fhall give a peculiar fenfation to one part, while it fhall give pain to another. Thus, what will produce ficknefs in the ftomach, will produce pain in the colon. When the fenfation of pain is in a vital part, it is fomewhat different from moft of thofe pains that are common. Thus, when the pain arifes from an injury done to the head, the fenfation is a heavy ftupifying pain, rendering the perfon affected unfit to pay attention to other fenfations, and is often attended with ficknefs, from the ftomach fympathizing with it.

When the pain is in the heart or lungs, it is more acute and is very much confined to the part difeafed.

When in the ftomach and inteftines, efpecially the upper part of them it is a heavy oppreffive fickly pain, but more or lefs, attended with ficknefs, according to its preffure or proximity to the ftomach ; for when fituated in the colon, it is more acute, and lefs attended with ficknefs.

We cannot give a better illuftration of this, than by taking notice of the effects of a purge. If we take fuch a purge as will produce both ficknefs and griping, we can eafily trace the progrefs of the medicine in the canal; when in the ftomach it makes us fick, but we foon find the ficknefs becoming more faint, by which we can judge that it has proceeded to the duodenum, and then a kind of uneafinefs, approaching to pain, fucceeds ; when this is the cafe, we may be certain that the medicine is paffing along the jejunum; it then begins to give a fickifh griping pain, which I conceive belongs to the ilium; and when in the colon it is a fharp pain, foon after which a motion takes place.

The liver, tefticles, and uterus, are fubject nearly to the fame kind of pain as the ftomach.

A tendon, ligament, and bone, give fomething of the fame kind of pain, though not fo oppreffive ; namely, a dull and heavy pain, often attended with fome little ficknefs, the ftomach generally fympathizing in fuch cafes.

But the fkin, mufcles, and the cellular membrane, in common, give an acute pain, which rather roufes than oppreffes, if not too great. All of this will be further mentioned when we treat of each part.

One caufe of this variety of fenfations, according as the parts inflamed, are vital or not vital, feems to confift in the different fyftems of materia vitæ with which thofe parts are fupplied, having, probably, nerves peculiarly conftructed for this purpofe; for all the parts which are fupplied with branches from the par vagum and intercoftals, effect the patient with lownefs of fpirits from the very firft attack of the inflammation: the actions of thofe parts are involuntary, and therefore are more immediately connected with the living principle, and confequently that principle is affected whenever any thing affects thefe nerves.

The other fyftem of the materia vitæ, when affected by this inflammation, rather roufes at firft the conftitution, which fhows figns of ftrength, unlefs the parts have rather weak powers of recovery, fuch as tendon, bone, etc. or

are far from the heart, in which cafes the figns of weaknefs, fooner or later, appear: hence it would feem that this
difference in the conftitution, arifing from the difference in
parts and their fituation, arifes from the conftitution having
a difeafe which it cannot fo eafily manage, as it can in thofe
parts which are not vital, and in parts that are near to the
heart, which circumftances alone become a caufe of irritation in the conftitution.

III. THE HEAT OF PARTS IN INFLAMMATION.

When I was treating of the blood, I obferved that the
heat of the animal was commonly confidered as connected
with that fluid; but as I had not made up my mind about
the caufe of the heat of animals, not being fatisfied with
the opinions hitherto given, I did not endeavour to offer
any account of that property; but I fhall now confider this
power when the animal is under difeafe, where it would
appear often to be diminifhed, and often increafed, and of
courfe the animal often becomes colder and hotter than its
natural temperature.

There is an endeavour to bring the heat a living body to
the temperature of the furrounding medium, but in the
more perfect animals this is prevented by the powers in
the animal to fupport its own temperature, more efpecially
in and near the vital parts ; therefore, in making experiments, to afcertain any variation, it is not neceffary to afcertain at the fame time the temperature of the atmofphere.

Heat, I imagine, is a fign of ftrength and power of conftitution, although it may often arife from an increafed
action either of weak conftitutions or of weakened parts.
Heat is a pofitive action, while cold is the reverfe, therefore producing weaknefs, and often arifing from a diminifhed action of ftrong parts.

It has not yet been confidered whether an animal has
the power of producing heat equally in every part of the
body; although from what is generally advanced on this
fubject, we are led to fuppofe that every part has this power;
or whether it is carried from fome one fource of heat by
the blood to every part; this may problably not be eafily

determined; but I am apt to fufpect there is a principal fource of heat, although it may not be in the blood itfelf, the blood being only affected by having its fource near the fource of heat.

That this principle refides in the ftomach is probable, or at leaft I am certain that affections of the ftomach will produce either heat or cold.

There are affections of the ftomach which produce the fenfation of heat in it, and the air that arifes in eructations, feels hot to the mouth of the perfon; but whether thefe fenfations arife from actual heat, or from fenfation only, I have not been able to determine.

Stimulating fubftances applied to the ftomach will produce a glow. Affections of the mind produce the fame effect, which laft circumftance might feem to contradict the idea of its arifing from the ftomach; but I fufpect that the ftomach fympathizes with thofe actions of the brain which form the mind, and then produces heat, which will be better illuftrated in treating of cold. I fufpect that the cold bath produces heat in the fame way from the fympathizing intercourfe between the fkin and the ftomach.

That difeafes augment or leffen this power in the animal is evident; for in many difeafes the animal becomes much hotter, and in many others much colder than is ufual to it. This fact was firft difcovered by fimple fenfation alone, both to the patients themfelves, and the practitioner, before the abfolute meafurement of the degrees of heat by inftruments was known; but it was impoffible that fuch knowledge of it could be accurate, for we find by experiment, that the meafurement of degrees of heat by fenfation is very vague. This happens becaufe the variations in the degrees of heat in ourfelves (which in fuch experimentsis the inftrument) is not of one ftandard, but muft vary pretty much before we are made fenfible of the difference, and therefore there can be only a relative knowledge refpecting our own heat at the time. But now our meafurement is more determined, and can be brought even nearer to the truth than is abfolutely neceffary to be known in difeafe.

The increafe and decreafe of the heat of an animal body may be divided into conftitutional and local. The conftitutional arifes from a conftitutional affection, and may arife primarily in the conftitution itfelf; or it probably may arife fecondarily, as from a local difeafe with which the conftitution fympathizes; but of this I am not yet certain, for from feveral experiments made to afcertain this point, it feemed to appear that local inflammation had little power of increafing the heat of the body beyond the natural ftandard, although the body was under the influence of the inflammation by fympathy, called the fymptomatic fever; but if the heat of the body is below the natural heat, or that heat where actions, whether natural or difeafed, are called forth, then the heat of the body it roufed to its natural ftandard*.

As it is the principle of increafe of local heat in inflammation I am now to confider, it fhould be firft afcertained how far fuch a principle exifts in a part, and what that principle may be; the conftitutional principle being in fome meafure not to the prefent purpofe, although it may throw fome light on the difference between the powers of the conftitutional, and thofe of the local principle. It is faid, that difeafe, as fever, has been known to raife the heat of the body to twelve degrees above the natural heat; and if fo, then there is in fuch cafes either an increafed power or an increafed exertion of that power; and to know whether this arifes only from a conftitutional affection at large, or whether it can take place in parts when the conftitution is affected by thofe parts, is worthy of inquiry.

The principal inftance of fuppofed increafed local heat is in inflammation; and we find that external parts inflamed do actually become hotter; but let us fee how far the increafe goes. From all the obfervations and experiments I have made, I do not find that a local inflammation can increafe the local heat above the natural heat of the animal; and when in parts whofe natural heat is inferior to that which is at the fource of the circulation, it does not rife fo high: thofe animals too, which appear to have no power either of increafe or decreafe in health, naturally appear to be equally deficient in difeafe ; as will be feen in the experiments.

I fufpect that the blood has an ultimate ftandard heat in itfelf, when in health, and that nothing can increafe that heat but fome univerfal or conftitutional affection; and probably the fympathetic fever is fuch as has no power in this way, and that the whole power of local inflammation

* Vide Animal Economy, page 87. vol. I.

is only to increase it a little in the part, but that it cannot bring it above the standard heat at the source, nor even up to it in parts that naturally or commonly do not come up to it, as just abovementioned.

As inflammation is the principal instance capable of producing local increased heat, I have taken the opportunity of examining inflammations, both when spontaneous and in consequence of operations. I have also made several experiments for that purpose, which are similar to operations, and cannot say that I ever saw, from all these experiments and observations, a case where the heat was really so much increased, as it appeared to be to the sensations.

EXPERIMENTS ON INTERNAL SURFACES.

EXPERIMENT I.

A man had the operation for the radical cure of the hydrocele, performed at St. George's Hospital. When I opened the tunica vaginalis, I immediately introduced the ball of the thermometer into it, and close by the side of the testicle. The mercury rose exactly to ninety-two degrees. The cavity was filled with lint, dipped in salve, that it might be taken out at will; the next day, when inflammation was come on, the dressings were taken out and the ball of the thermometer introduced as before, when it arose to ninety-eight degrees and three-fourths exactly.

Here was an increase of heat of six degrees and three-fourths; but even this was not equal to that of the blood, probably, at the source of the circulation in the same man. This experiment I have repeated more than once, and with nearly the same event.

As the human subject cannot always furnish us with opportunities of ascertaining the fact, and it is often impossible to make experiments when proper cases occur, I was led to make such experiments on animals, as appeared to me proper for determining the fact; but in none of them could I ever increase the inflammatory heat so as to make it equal to the natural heat of the blood at its source.

EXPERIMENT II.

I made an incision into the thorax of a dog, the wound was made about the centre of the right side, and the thermometer pushed down, so as to come in contact, or nearly

C 2

fo, with the diaphragm. The degree of heat was one
hundred and one; a large doſſile of lint was put into the
wound to prevent its healing by the firſt intention, and covered over by a ſticking plaſter. The dog was affected with
a ſhivering. The day following the lint was extracted and
the thermometer again introduced, the degree of heat appeared exactly the ſame, viz. one hundred and one. This
dog recovered.

<h3 style="text-align:center">EXPERIMENT III.</h3>

An oblique inciſion was made about two inches deep into
the gluteal muſcles of an aſs, and into this wound was introduced a tin canula about an inch and half long, ſo that half
an inch of the bottom of the wound projected beyond the
canula; into this canula was introduced a wooden plug,
which projected half an inch beyond the canula, ſo as to fill
up the bottom of the wound, and which kept that part of
the wound from uniting. The whole was faſtened into
the wound by threads attached to the ſkin.

Immediately upon making the wound the ball of the
thermometer was introduced into it to the bottom, and the
mercury roſe to one hundred degrees exactly, as it did alſo
at the ſame time in the vagina.

On the next morning the wooden plug was taken out, and
the ball of the thermometer (being previouſly warmed to
ninety-nine degrees) was introduced down to the bottom
of the wound, which projected beyond the canula, and the
mercury roſe to one hundred degrees. The wooden plug
was returned and ſecured as before, In the evening the ſame
experiment was repeated, and the mercury roſe to one
hundred degrees. Friday morning it roſe only to ninety-
nine degrees. Friday evening it roſe to near one hundred
and one degrees and half. Saturday morning, ninety-nine
degrees, and in the evening one hundred degrees.

A ſimilar experiment to this was made on a dog, and the
heat was one hundred and one degrees. The day following
the heat was the ſame, as alſo on the third day, when ſuppuration was taking place.

<h3 style="text-align:center">EXPERIMENT IV.</h3>

Although in the experiment upon the dog, by making an
opening into the thorax ſo as to excite an inflammtion there,
and to affect his conſtitution, the heat of the part was not

increafed; yet in order to be more clear with regard to the
refult of fuch an experiment, a wound was made into the
abdomen of an afs, and a folution of common falt and water
thrown in (about a handful to a pint of water), to excite an
univerfal inflammation in the cavity of the abdomen. This
produced great pain and uneafinefs, fo as to make the animal
lie down and roll, becoming as reftlefs as horfes when griped.

The next morning, Friday, the thermometer was introduced into the vagina, and the mercury ftood at ninety-nine
degrees and a half, nearly the fame heat as before the experiment; in the evening one hundred and one degrees and
a half. Saturday morning, one hundred degrees and a
half ; evening, a hundred degrees and a half. The vagina
therefore, was not rendered hotter by an inflammation which
produced what we may call the fympathetic fever.

The animal was now killed, and on examining the abdomen, the fide where the wound was made appeared much
inflamed, as well as the inteftine oppofite to this part. All
of them adhered together, and the inteftine furrounding this
part of the adhefions had their peritoneal coat become extremely vafcular, and matter was formed in the abdomen.

But that the heat of a part can be increafed above the
common ftandard of a healthy perfon is certain, when it is
fuch a part as is naturally of the ftandard heat; as for inftance, the abdomen. For in lord Hertford's fervant, who
was tapped eight times, and feven of them in thirteen weeks,
the feventh time I held the ball of a thermometer in the
ftream, as it flowed from the canula of the trochar, and it
raifed the mercury to one hundred and one degrees, exactly,
through the whole time. Twelve days after I tapped him
the eighth time, the water was pretty clear; when I held the
thermometer in the ftream, it rofe to one hundred and four
degrees. Now as the heat of the abdomen was one hundred
and four, we muft, I think, fuppofe that the general heat of
the man would alfo be one hundred and four degrees.

EXPERIMENTS ON SECRETING SURFACES.
EXPERIMENT I.

I took the degree of heat of a dog's rectum, by introdu-
cing the thermometer about three inches; and when it was
afcertained, four grains of corrofive fublimate were diffolved in two ounces of water, and the folution thrown up the
rectum. The day following the thermometer was again
introduced, and then I found the heat fomewhat increafed,

but not quite a degree. As far as one might judge from external appearances, the rectum was very much inflamed, as there was a considerable external swelling, forming a thick elevated ring round the anus.

EXPERIMENT II.

I introduced into the rectum of an afs, the thermometer, and the mercury rose to ninety-eight degrees and a half exactly : this was repeated several times with the same result. I then threw up the rectum an injection of flour of mustard and ginger, mixed in about a pint of water. About twelve hours after, I introduced the thermometer, and it rose to ninety-nine degrees and an half.

The injection was repeated several times, but the heat did not increase.

EXPERIMENT III.

To irritate the rectum still more, I threw up a solulution of corrosive sublimate ; and about twelve hours after, I introduced the thermometer, and found no increase of heat. Twenty hours after, I introduced the thermometer ; but the heat was the same. Sixty hours after the injection, the thermometer being introduced, the mercury rose to one hundred degrees, exactly. This injection had irritated it so much, as to give a very severe tenesmus ; and even blood passed.

EXPERIMENT IV.

The natural heat of the vagina of a young afs was one hundred degrees. A solution of corrosive sublimate, as much as would dissolve in a tea-cup full of water, viz. about ten grains, was injected into the vagina. In about two hours after, the mercury fell to ninety-nine degrees. Thursday morning ninety-nine degrees, evening one hundred. Friday morning ninety-nine, evening near to one hundred and one. Saturday morning ninety-nine, evening one hundred degrees.

This experiment was repeated several times upon the same afs, with the same result.

In these experiments it can hardly be said, that the heat was increased. That the inflammation had been raised to a very considerable degree was plain, for it produced a discharge of matter which was often bloody, and upon kill-

ing the afs for another experiment, the following appearances were found in the uterus.

The horns of the uterus were filled with ferum, and the
inflammation had run fo high by the ftimulating injections which were ufed for the experiments on the vagina,
that the coagulating lymph had been thrown out fo as almoft to obliterate the vagina, uterus, etc. by thofe adhefions which are the ultimate effects of inflammation on fecreting canals, while fuppuration is the ultimate effect of
inflammation on internal furfaces : there were no figns
of inflammation on the external furface of the uterus
which is covered by the peritoneum.

It may juft be remarked, that in moft of thofe experiments the heat in the morning was a degree lefs than in
the evening ; and I may alfo remark, that this is commonly the cafe in the natural heat of the animal.

I wifhed to know whether fuch animals as have little
or no power of varying their natural heat, had a power of
increafing their heat in confequence of injuries; for which
purpofe I opened into circumfcribed cavities in frogs,
toads, and fnails, and at different periods, after the opening was made, the thermometer was introduced. As the
heat of thofe animals is principally from the atmofphere,
the external heat is to be connected with the experiments.

NOVEMBER 27, 1788.

A healthy toad and frog, after having the heat in the
ftomach afcertained, had openings made through the fkin
of the belly, large enough to admit a thermometer, and
the orifice was kept open by a piece of fponge.

Atmofphere . . . . 36°
Stomach of both . . . . . . 40°
Under fkin of the belly . . . 40°

	Atm.	Frog.	Toad.	Stom.
		Under the fkin.		
Half an hour after the opening	35°	40°	40°	40°
Hour and a half	35°	39°	39°	
Two hours and a half		39°	39°	

The abdomen was now opened, and a piece of fponge
kept in the orifice.

	Atm.	Frog.	Toad.	Stom. Abdomen.
The heat	36°	40°	40°	40°
Hour and a half after opening . .	36°	39°	39°	39°
Four hours and a half . . .	38°	39°	39°	

Part of the left oviduct protruded of the natural colour and appearance.

	Atm.	Frog.	Toad.	Stom.
Twenty-one hours and a half	38°	38°	38°	38°
Nine hours after	35°	35°	35°	35°

The protruded oviduct was more vascular and of a uniformly red appearance ; it was returned into the belly and retained there.

	Atm.	Frog.	Toad.	Stom.
Twenty-four hours	32°	32°	32°	32°
Forty-six hours	34°	34°	34°	34°

The toad died, and the frog was become very weak and languid : part of the oviduct protruded and had the small vessels loaded with blood.

It lived one hundred and eighteen hours, that is, seventy-two longer than the toad, during which period its heat corresponded with the atmosphere.

Upon examining the abdomen after death, there were no adhesions nor any appearances from inflammation, except on the protruded oviduct.

Some healthy shell-snails had openings made into the the lungs, and their heat ascertained at the following times.

	Atm.	Snail.
The heat at the time . . .	34°	38°
One hour and a half . . .	32°	32°
Six hours and a half . . .	32°	35°
Ten hours	31°	36°
Twenty-four hours	30°	30°

To ascertain the standard heat of a snail.

	Atm.	Snail.
A fresh lively snail had its heat in the lungs	30°	. . 36°
Another	28°	. . 35°
Another	30°	. . 37°

EXPERIMENTS to afcertain the heat of worms, leeches
and fnails, when compared with the atmofphere, and the
changes produced in their heat by inflammation.

EXPERIMENT I.

Heat of the air in the room . . 56°
———— water in the room . . 57°
———— fome earth-worms . . 58° $\frac{1}{2}$

EXPERIMENT II.

Water as a ftandard . . . . 56° $\frac{1}{4}$
Leeches in the fame quantity . 47°

EXPERIMENT III.

Water as a ftandard . . . . 56°
Frefh egg . . . . 55°
Leeches alone . . . . 60°
Worms alone . . . . 57°
Air . . . . 54°
Worms ⎫ . . . . 58°
Leeches ⎬ two hours after being wounded . 57°
Slugs ⎭ . . . . 58°
Air . . . . 55°
Worms ⎫ . . . . 55°
Leeches ⎬ twenty-four hours after being wounded 55°
Slugs ⎭ . . . . 55°
They were all very weak and dying.

IV. OF THE PRODUCTION OF COLD IN IN-FLAMMATION.

THE production of cold is certainly an operation which
the more perfect animals are endowed with; and this power would appear to be both conftitutional and local, fimilar
to the power of producing heat. As the word inflamma-

Vol. II.   D

tion implies heat, and has been used to exprefs that action of the veſſels where heat is commonly an effect, it may ſeem ſtrange that we ſhould treat of cold in the action of inflammation ; but probably we have no action in the body that is not attended with an occaſional production of cold ; how far this takes place in parts I do not know ; but that it takes place conſtitutionally, from almoſt every affection, is evident, whether it be inflammatory fever, or local inflammation. As an animal has no ſtandard of cold, but at the ſource, which is alſo the ſtandard of heat, it is perhaps impoſſible to aſcertain with certainty the degree of cold produced either by diſeaſe, or from the ſurrounding cold ; but perhaps by comparing the part ſuſpected of being colder than is natural from diſeaſe, with a ſimilar part under the ſame external influence of heat and cold, as for inſtance, one limb with the other, or one hand with the other, a pretty fair inference may be made ; and we often find that diſeaſed parts ſhall become extremely cold, while from other circumſtances than diſeaſe they ſhould not be ſo.

I ſuſpect that coldneſs in diſeaſe ariſes either from weakneſs, or a feel or conſciouſneſs of weakneſs in the whole conſtitution or a part, joined with a peculiar mode of action at the time.

Thus we have many conſtitutional diſeaſes beginning with abſolute coldneſs, which ſeems afterwards to terminate in a ſenſative coldneſs only, as the cold fit of an ague; for I apprehend that the ſickneſs which generally proceeds ſuch complaints, produces univerſal cold, and once having produced the action of the body ariſing from abſolute cold, the action goes on for ſome time, although the cauſe no longer exiſts, which continues the ſenſation ; and although the abſolute coldneſs is gone, yet the action of the parts, which is a continuation of, and therefore ſimilar to the action of the abſolute cold, is capable of deſtroying itſelf by producing the hot fit, if there be power or diſpoſition.

That weakneſs, or a feel of weakneſs, produces cold is evident ; and that univerſal or conſtitutional cold ariſes from the ſtomach is alſo evident ; for whenever we are made ſick an univerſal coldneſs takes place ; and this is beſt proved by producing ſickneſs on animals that we can kill, or that die while they are under theſe affections of the ſtomach. The experiments I made to aſcertain this were not conducted with great accuracy, as I trusted in them entirely to my own sensations or feelings.

EXPERIMENT. I threw three grains of tartar emetic into the veins of a healthy bitch, the quantity of water near an ounce. In about twenty minutes she had a stool and voided some single tape-worms. Some of the stools were extremely thin, and made up principally of bile. Some time after she had two more stools, which were thin and bilious. She continued pretty easy for about three hours, but became a little convulsed, which increased, and at last she became senseless, with little twitchings; hardly breathing, except with the diaphragm, and having a low, slow pulse. She was very cold to our feel, when applying our hands on the skin of the body. In about ten or twelve hours after the injection, she died.

EXPERIMENT. I repeated the above experiment on another bitch, adding a full grain more to the medicine. She vomited in less than a minute after it was thrown in, and strained excessively hard, throwing up a great deal of froth, which was only the mucus of the stomach mixed up with the air in the act of reaching. In less than three minutes she had a stool, which was pretty loose and partly of the natural appearance. She continued reaching and purging for above an hour, and was extremely uneasy, at last she got into a dark corner and lay there, frothing at the mouth, was taken with convulsive twitchings like the former, and died in about five hours after the injection. I opened her body immediately after death, and found the intestines, liver, and heart, not so warm as we usually find them.

I have known people who had affections of the stomach and bowels, say, that they had plainly the feeling of cold in their bellies. I knew a gentleman who told me, that often when he threw the wind off his stomach, it felt cold to his mouth and even to his hands, which was by much the best guide respecting sensation.

A lady near seventy years of age, has a violent cough, which often makes her puke, and what comes off her stomach feels like ice to the mouth.

Affections of the mind also produce constitutional coldness, but they are such affections as the stomach sympathizes with, producing sickness, shuddering, etc.

A disagreable idea or sight will sometimes give a quick

fenfation of ficknefs, and the fkin fhall fympathize with
the ftomach, it fhall appear to begin, as it were, in the
mouth or throat, as if fomething there had a tendency to
come up ; the mufcles of the neck fhall become convulfed,
and the head fhall be violently fhaken ; from thence a difagreeable feeling fhall fpread over the whole body, paffing
directly down the back to the feet ; commonly expreffed
by faying " ones flefh creeps" ; and hence the words, fhud-
der and horror, exprefs mental as well as bodily affections.
Another action fhall be joined with the cold, viz. the action of fweating, fo that a cold fweat fhall take place over
the whole body. This cold fhall be partial, for under many
difeafes a partial cold fweat will fometimes come on, while
other parts remain tolerably temperate.

V. OF THE TIME THE ADHESIVE INFLAM-MATION COMMENCES AFTER ITS CAUSE; AND IN WHAT CASES AND PARTS IT IS IMPERFECT IN ITS CONSEQUENCES.

IT will be often impoffible to determine the diftance of
time between the impreffion which becomes the caufe of
inflammation, and the action itself, which will depend
upon two circumftances, viz. the nature of the exciting
caufe, and the fufceptibility for fuch action in the parts.

In the expofure of internal furfaces, inflammation is perhaps fooner brought on than in moft others ; for the
incitement is immediate, and there is no remiffion in the
caufe itfelf.

In fpecific difeafes its time is perhaps more regular, each
having a determined interval between the application of
the exciting caufe and the appearance of the difeafe, although even in fome of thefe there is a vaft difference in
the time after contamination, but in thofe arifing fpontaneoufly it muft be uncertain ; yet in fome cafes it can be
pretty well afcertained, fuppofing fenfation the firft effect
of the inflammatory impreffion ; and in fuch inftance we
often find it very rapid. They fhall be attacked with a
violent pain in the part, fo much fo as hardly to be able to

bear it, which shall be immediately succeeded with a violent inflammation.

A lady was walking in her garden, and at once was attacked with a violent pain in the middle of the fore part of the thigh, which made her immediately lame; soon after, the skin appeared difcoloured, which spread nearly over one-half of the thigh; this part became thick and swelled, which appeared to go as deep as the bone; it afterwards suppurated, all in a few days; this appeared to be a well-marked cafe.

The commencement of inflammation after accidents is more easily ascertained, for we muft date it from the accident, and we find it is not immediate; for after a wound has been received, inflammation does not begin for twelve, eighteen, or twenty-four hours.

It sometimes happens, however, that the adhesive state cannot set bounds to itself, and therefore cannot set bounds to the suppurative. This may be owing to two caufes; the one is, the violence of the inflammation, and quicknefs of the attack of the suppurative spreading before parts have had a sufficient union, and even perhaps joined with a species of suppuration from the very firft, fo that union is prevented. Secondly, the inflammation may, I fufpect, be of the eryfepelatous kind, efpecially when there is a tendency, from the beginning, to mortification.

This mixing of the suppurative with the adhesive, or the hurrying on of the fuppurative, or this mixture of the eryfepelatous with the others, I have frequently seen in the abdomen of women who have been attacked with the peritoneal inflammation after child-birth, and which from thefe circumftances became the caufe of their death.

In fuch cafes we find matter mixed with coagulating lymph, as if formed with it; for without having been formed with it, it could not have mixed with it after coagulation; we find alfo coagulating lymph mixed with the matter, as it were, feparated from the inflamed furface by the formation of the matter; and in thofe cafes where there is a tendency to mortification from the beginning, as in ftrangulated ruptures, we often find the adhesive and fuppurative inflammation going hand in hand. All of thefe caufes and effects account for the violence of the symptoms, the quicknefs of the progrefs of the difeafe, and its

fatal confequences beyond fuch inflammations as have only the true adhefive progrefs, or where it takes place perfectly prior to fuppuration.

It feems to appear from obfervation, that fome furfaces
of the body do not fo readily unite by the coagulating
lymph as others, and therefore, on fuch furfaces there is
commonly a much larger quantity of this matter thrown
out than probably would have been if union had readily
taken place ; for we may fuppofe, that where once union
has taken place, extravafation is at an end.   Thus, we fee
in (what we may fuppofe) inflammation of the heart, that
the coagulating lymph is thrown out on the exterior furface in vaft quantities, while at the fame time the heart
fhall not adhere to the pericardium.   This is not only feen
in the human, but in other animals.   In an ox, the heart
was furred all over, and in fome places, the coagulating
lymph was near an inch in thicknefs.   The external furface of fuch hearts have an uncommon appearance ;   the
outer furface of the coagulating lymph is extremely irregular, appearing very much like the external furface
of a fponge, while the bafe, or attachment to the heart
is very folid and firm.   However, in many inftances we
find the pericardium adhering to the heart, and generally in pretty clofe contact, which would make us fuppofe that the extent of motion of thofe two parts on one
another is not great.   Thefe adhefions affect the pulfe
much, which is a good reafon why nature avoids them
as much as poffible.   On the other hand, it feems de-
ducible from obfervation, that neither the pia nor dura-
mater are apt to throw out much coagulating lymph, for
here it would produce compreffion ; and, therefore, we
feldom find adhefions between them ; in confequence of
fuch accidents as produce fuppuration between thefe two
membranes, we feldom or ever find the furrounding
parts adhering fo as to confine the matter to the fuppurating furface.

Inflammation of the fkin, or fuch as approaches to the
fkin produces in general a feparation of the cuticle, often
of the hair, or the nails.   Thefe effects arife fooner or later, according to the nature and degree of the inflammation, but more particularly according to its nature ; they
take place the leaft and lateft in the true adhefive inflammation, which is always attended with the greateft ftrength.

In fuch cafes, the feparation does not happen till the inflammation has fubfided ; and as a proof of this, in the
gout, it is leaft and lateft of all ; for this is always a healthy
inflammation, otherwife it would not take place ; but in
weak habits, at the early part of the difeafe, there are often vefications, which are filled with ferum, fometimes
with coagulating lymph, etc. both of thefe are fometimes
tinged with red blood; when the inflammation is of a weak
kind, tending to mortification, the cuticle commonly fe-
parates early during the time of inflammation, almoft beginning with it, and of courfe the vefications will be filled with ferum, and often with the red globules ; we may
obferve in wounds of the fkin which are not allowed to
heal by the firft intention, that a feparation of the cuticle
will take place at the edges of the wound, and this will extend according to the nature of the inflammation, which
is according to the nature of the conftitution ; this will be
attended with other concomitant appearances, fuch as flab-
by edges and thin matter : I conceive, in the weak habit it
depends on an action of the inflammation itfelf ; but in the
ftrong, it depends on a ftate in which the parts are left to
feparate the cuticle.

This feparation arifes, I apprehend, from a degree of
weaknefs approaching to a kind of death in the connection
between the cuticle and cutis, from life being in this part
naturally very weak. In the beginning of mortification it
is produced ; in the œdematous and eryfepelatous inflammations it is greateft, and in putrefaction of dead bodies
it is the firft operation. I fufpect too, that a bliftering
plafter, hot water, etc. only kills the uniting parts, by
which means an irritation is produced in the cutis, and the
extravafation is according to that irritation.

The connection of the cuticle is more or lefs deftroyed
in every inflammation of the fkin ; for we feldom fee an
inflammation attack the fkin but the cuticle comes off fooner or later ; we generally obferve it peeling off in fleakes,
after inflammation has fubfided, and it begins neareft the
point of inflammation*.

* It may be obferved, that when an inflammation attacks
the finger ends, or toes, fo as to produce fuppuration either in
the fubftance of thefe parts, although not larger than a pimple,
or only on the furface of the cutis, an extenfive feparation of
the cuticle takes place, not entirely from the inflammation,

VI. OF THE UNITING MEDIUM IN INLFAM-MATIONS.

EVERY new fubftance that is formed is either for a fa-lutary purpofe, or it is difeafed. The firft confifts either of granulations, or of adhefions, whether with the firft, or fe-cond intention ; and all thefe may be confidered as a re-vival of the rational principles and powers of growth, whereas difeafed fubftances are, as it were, monfters.

In the adhefive inflammation, the veffels being enlarged, as above defcribed, fimilar to what they are in the young fubject, begin to feparate from the mafs fome portion of the coagulating lymph, with fome ferum, and alfo red globules, and throw it out on the internal furface ; probably through the exhaling veffels, or perhaps, open new ones, and cover the fides of thofe cells, which eafily unite with the oppofite, with which they are in contact, forming the firft progrefs of adhefions.

That this is really the cafe, and that this effect has taken place in confequence of inflammation, is evident from the following obfervations. In all large cavities, where we can make our obfervations with certainty, when in the ftate of inflammation, we find diffufed over the fides, or through the cavity, a fubftance exactly fimilar to the coagulating lymph when feparated from the ferum, and red blood, after common bleeding. That the blood, when thrown out of the circulation from an inflammatory ftate of the veffels, as well as the blood itfelf, unites parts together, is probably beft feen in the inflammation of the larger cavities above-mentioned. The following I fhall give as an example, which I have often obferved on the peritoneum of thofe who have died in confequence of inflammation of this membrane. The inteftines are more or lefs united to one another, and,

but affifted by it : this is owing, principally, to the cuticle in fuch places not giving way, being there ftrong, fo that a feem-ing abfcefs almoft occupies the whole finger, etc. this fhould be opened early to prevent this feparation as much as poffible, or to prevent the feparation from extending too far.

according to the stage of the inflammation, this union is stronger or weaker ; in some it is so strong as to admit of some force to pull them asunder ;* the smooth peritoneal coat is, as it were, lost, having become cellular, like cellular membrane. When the vessels of this part are injected we shall find, that in those parts where a separation has been made by laceration, previous to the injecting, the injection will appear on that surface like small spots or drops, which shews that the vessels had at least passed to the very surface of the intestines.

In parts where the union was preserved, I have observed the three following facts. On separating the united parts, I have observed, in some places, the vessels come to the surface of the intestines, and then terminate all at once. In other places, I could observe the vessels passing from the intestine into the extravasated substance, and there ramifying, so that the vessel was plainly continued from the old into the new.

In a vast number of instances, I have observed, that in the substance of the extravasation, there were a great number of spots of red blood in it, so that it looked mottled. The same appearance was very observable on the surface of separation, between the old substance and the new, a good deal like petechial spots. How this red blood got here is the thing to be considered, especially as a good deal was within the substance of the coagulum. Was it extravasated along with the coagulating lymph ? In this case, I should have rather supposed it would have been more diffused, and if not diffused, more attached to the intestines, and not in the centre of the coagulum ; if it had been extravasation, one would have expected extravasation of injection, but we had none in any of these places ; I have therefore suspected, that parts have the power of making vessels and red blood independent of circulation. This appears to be evidently the case with the chick in the egg.

* Adhesions in consequence of inflammations become very soon strong, and are very soon elongated ; probably as soon as they become organized they adapt themselves to their situation or the necessity. Thus the dog who had his belly opened to wound some lacteals, when killed on the ninth day, had his intestines connected by adhesion in several places, and those very firm and long.

I have obferved, when I was treating of the blood, that
it was capable of becoming vafcular, when depofited either
by accident, or for particular purpofes ; and I had reafon
to believe, that a coagulum, or a coagulating lymph had a
power of becoming vafcular in itfelf, when it could be fupplied with blood, and mentioned the coagulum in a larger
artery as an inftance. Likewife when I was treating of
union by the firft intention, I explained the intercourfe
eftablifhed by the uniting medium becoming vafcular, and
thofe veffels uniting acrofs by a procefs, called inofculation.
The fame reafoning is applicable to the union by means of
the adhefive inflammation ; for it is the blood in all cafes
that is to become vafcular; but this takes place fooner or
later, according to the apparent neceffity. In fome it becomes vafcular, immediately ; in others very late ; and indeed, in fome hardly ever, according to the degree of utility to arife from that change. Where it becomes vafcular
fooneft, there the veffels are found alfo in greateft numbers,
the two effects depending on the fame principle.

Extravafation, whether of blood, or only of lymph, becomes vafcular, almoft immediately, when thrown out into
the cavity of the human uterus into the ftate of pregnancy.
Here is an operation neceffary to go on, which is more than
the fimple fupport of the extravafation itfelf ; but when the
extravafation is thrown out by accident, or for the purpofe
of producing adhefions, the immediate intent is anfwered
without the veffels, and vafcularity only becomes neceffary
afterwards ; therefore vafcularity in fuch cafes is the fecond
confideration, not an immediate one. But in the cafe of
impregnation it muft be immediate, for the fimple extravafation would not anfwer the intention. This fhews that
this extravafation is very different from that of the menfes.

The new veffels which are formed in the newly extrava-
fating and uniting fubftance, become of ufe both during
the ftate of adhefion and fuppuration.

In the firft, they ferve to give powers of action to this
new fubftance, which affifts in preventing fuppuration. In
the fecond, where this cannot be done, they affift in forming a vafcular bafis for the granulations.

When we cut into inflamed parts after death, we find
them firm and folid, refembling the fection of a lemon, or
fome œdematous tumor, where we know extravafation has
taken place.

This appearance arifes from the cells in the cellular membrane, and other interftices of parts, being loaded with extravafated coagulating lymph ; from this circumftance they are cemented together and become empervious to air, not fimilar in thefe refpects to common cellular membrane, or natural parts. In many places where this extravafation has been in confiderable quantities, it is formed in time into cellular membrane.

I have obferved, that this mode of the feparation of coagulating lymph is not peculiar to inflammation ; it is feparated in many difeafes.

It is thrown out to form tumours, etc. where inflammation does not feem to be a leading caufe ; and we often find the adhefive ftages, as it were, degenerate into, or terminate in the formation of a cyft, to contain the body that was the caufe of the inflammation. Thus, a fack is formed for bullets, pieces of glafs, etc.

It is unneceffary to inftance every poffible fituation where adhefions could be produced ; they can take place wherever there two internal furfaces in contact, or that can be brought into contact. I cannot give a better inftance of its utility in the animal œconomy than in the following experiment : I wifhed to know in wounds which penetrated into the cheft, (many of which I have feen in the army) where fuppuration had come on the whole cavity of the cheft, as well as on the furface of the lungs, and where the lungs collapfed, how parts were reinftated, or in what form they healed ; whether the lungs, etc. loft their fuppurating difpofition, and dilated, fo as to fill the cheft again. To afcertain this as far as one well could, I made the following experiment on a dog.

October 1777, I made an opening between the ribs into the cheft of a dog, and touched the edges of the wound all round with cauftic to prevent it from healing by the firft intention, and then allowed the dog to do as he pleafed. The air at firft paffed in and out of his cheft by the wound. He eat, etc. for fome days, but his appetite gradually began to fall off. He breathed with difficulty, which increafed ; he lay principally on that fide which we find people do who have the lungs difeafed in one fide only or principally ; and he died the eleventh day after the opening. On opening the body, I found the collapfed lungs paffing directly acrofs the cheft and attached to the infide of the wound all round so that they excluded the cavity of the cheft from all external communication. This circumftance of the lungs falling acrofs the cheft was owing to his having lain principally on that fide, which I conceived to have been only accidental.

The cavity of the cheft all round was filled with air. That part of the external furface of the lungs which did not adhere, that is to fay, the upper furface of the diaphragm, and that part of the pleura which covered the ribs were entirely free from inflammation, or fuppuration ; this cavity, from thefe adhefions, being rendered a perfect cavity, fhews that air, fimply has no power to excite inflammation when the cavity is otherwife perfect; which the adhefions had effected ; this fhews alfo that adhefions of two furfaces round the expofed part, exclude every part from the neceffity of inflammation, as was explained when treating of inflammation.

From the connection between the living powers of the folids and the fluids, we can hardly fuppofe that fuch an uncommon action could take place in the vafcular fyftem, without producing its effects upon the fluids ; and, therefore, from reafoning we might fuppofe, that the coagulating lymph undergoes fome changes in its paffage through the inflamed veffels, which obliges it to coagulate more immediately, or much fooner than it otherwife would†.

For in thofe cafes of inflamed arms, after bleeding, and inflammations in confequence of other caufes, we find that the cavities of the veins are in many places furred over, and in others united by means of the coagulating lymph. Now if this coagulating lymph is fimilar in its productions to that which we have been defcribing, it muft have been thrown out from the vafa vaforum, thefe veffels having feparated it and poured it into the cavity of the veins, and it muft there have coagulated immediately ; in this feparation therefore, from the blood, it muft have undergone fome change, arifing from the actions of the veffels ; for if this

† This is contrary to the difpofition of inflammatory blood when taken out of the veffels and allowed to go through its fpontaneous changes ; from which it would appear that the general affection of the blood (which I would call fympathy of the coagulating lymph) with the univerfal irritation) is different from its affection or difpofition when employed for the purpofes of union

lymph was no more than the coagulating lymph with its common properties, or the properties common to that which is circulating in the same vein which receives it, it would in such cases only continue to throw in more coagulating lymph, in addition to what was circulating, and therefore, probably, it would be carried along with the blood to the heart, as a part of the common mass. From this we should infer that this coagulating matter is not simply the coagulating lymph, such as it is when circulating, but somewhat different, from having undergone a change in its passage through the inflamed vessels, partaking of the disposition of those solids which are inflamed, through which it passed. This process cannot, therefore, be supposed to be merely extravasation ; for I conceive that an œdema would be a consequence of simple extravasation. But this may be taken up in another point of view, and upon the same principle. The inflamed vessels may give a disposition to the blood, as it is moving slowly along, to coagulate on its surface ; and this is probably the more just idea of the two ; as we find that the vessels, both veins and arteries, can give this disposition, and to a very great extent : we find in the beginning of mortification, the blood coagulating in the vessels, so as to fill them up entirely, and this preceeding the mortification, seems to be for the purpose of securing the vessel before it is to give way; we, therefore, cannot doubt of a coagulating principle being given to the blood from the vessels; and as a further proof of this, we may observe that the extravasated coagulating lymph, which produces either adhesions or forms tumors, (which is often the case) is always of the nature of the diseased solids that produced it. If the case is venereal, the new substance is of the same nature ; if cancerous, it is cancerous; for I find that it has, when absorbed, the power of contaminating, similar to matter or pus produced by the sores or ulcers of such diseases ; the absorbent glands being often affected by the absorption of the coagulating matter of a schirrous beast.

Whatever change the coagulating lymph has undergone in this operation of inflammation, it seems so far the same, as to retain still the nature of the coagulating lymph, and to possess the living principle ; this is most probably in a greater degree, and therefore, the coagulating lymph is still better fitted to be formed into a part of the solids of the bo-

dy, as will be taken notice of when we come to treat of the
ftate of the blood in the inflammation.

But it is not abfolutely neceffary that the coagulating
lymph fhould firft undergo a change in the extravafated veffels, before it can become a living folid, or unite living folids; for we find that common blood extravafated from a
ruptured veffel is, perhaps, equally efficacious in this refpect,
therefore the red globules do not retard union, but they
may promote it.

VII. THE STATE OF THE BLOOD, AND OF THE PULSE, IN INFLAMMATION.

FROM what has been faid of the living power of the blood
I think we muft allow that it will be commonly affected
much in the fame manner with the conftitution, and that
difeafe will have nearly the fame effect upon it, as it has on
the body; becaufe, the fame living principle runs through
the whole. We find this to be nearly the cafe; for till a
difeafe has affected the conftitution, the blood continues
the fame as before; but as the conftitution becomes affected, the blood alfo becomes affected, and undergoes the
fame changes, which, probably, may be afcribed to contiguous fympathy between the veffels and the blood; and we
fhall find that the changes in the blood is often as much expreffive of difeafe as any other part of the body. It is expreffive of ftrong action, as well as of weak action; but as
it does not give fenfation, it cannot convey to the mind all
the varieties of difeafe that may take place in it; yet I could
conceive, if the blood was to be primarily affected, that an
impreffion would be made upon the mind, from its affecting
the veffels in which it moved. However, it is not always
the cafe that the ftate of the blood and the other fymptoms
are expreffive exactly of the fame thing; the blood often
expreffing lefs, and often more; when the action of the folids is of the inflammatory kind, or which, perhaps, is the
fame thing, when there is too great an action of the folids,
the blood more readily admits of a feparation of its vifible

parts, and the coagulating lymph' coagulates more flowly, but becomes firmer when coagulated ; this laft circumftance however, might be fuppofed not to be fo clear, for its firm-nefs may be owing to its want of the red particles, which certainly gives the blood a brittlenefs in proportion to their quantity ; but although this may have fome effect, yet it is very little ; for we find blood of loofe texture in fome in-flammations, when deprived of its red part ; when blood has this difpofition it is called fizy blood. Thefe changes in the nature of the blood depend fo much upon the above-mentioned caufes of inflammation, that it is impoffible to fay whether they do not conftitute the firft univerfal effect produced from the local inflammation, and whether the conftitutional is an effect of this change in the blood. I knew a man who was ftabbed in the loins, and according to the confequent fymptoms, was moft probably wounded or hurt in fome vifcus within the abdomen. At firft he had no fymptoms, but fimple pain in the part, I therefore only bled him, by way of precaution, and the blood was perfect-ly natural ; in lefs than a quarter of an hour after, conftitu-tional fymptoms came on, fuch as rigor, ficknefs, etc. and on opening the fame orifice; and taking away more blood, this fecond quantity had a very thick and ftrong buff upon it, having all the appearance of inflammatory blood ; while this conftitutional difpofition lafted, which was fome time; his blood continued the fame, which was proved by the fubfe-quent bleedings. The fubfiding, however, of the red glo-bules in the blood when in an inflamed ftate, although pret-ty frequent, is not always an attendant, or in other words (and perhaps upon fome other principle) the blood is not al-ways attended with this appearance, when the vifible fymp-toms are the fame. A young woman was attacked with a violent cough, oppreffion in breathing, quick, full, and hard pulfe. She was bled, which gave her eafe; the blood was fizy; the fymptoms again returned, and fhe was bled a fe-cond time ; which alfo relieved her, and the blood was more fizy than before, fo far all the fymptoms agreed ; the fymptoms again recurred and were more violent than be-fore ; fhe was bled a third time, and a third time relieved; but this blood was not in the leaft fizy, although it came from the vein very freely. Here then, the blood, under the fame difeafe, loft this difpofition, although the fymp-toms remained the fame. As inflamed blood leaves a por-

tion of the coagulating lymph free from the red globules at
the top, and as that can be accounted for upon the principle
of the coagulating lymph, in fuch cafes not coagulating fo
faft as when the blood has not this appearance, and as the
coagulation hinders any comparative experiment refpecting
the weight of the red globules of each, I tried to fee if they
funk in ferum fafter in the one kind of blood than in the o-
ther ; I took the ferum of inflammatory blood, with fome
of the red part, and alfo fome ferum of blood free from inflammation, with nearly the fame quantity of the red part;
they were put into phials of the fame fize ; I fhook
them at the fame time, then allowed them to ftand quiet,
and obferved that the red globules fubfided much fafter in
the inflammatory blood than in the other. To afcertain
whether this arofe from the red globules being heavier, or
the ferum lighter, I poured off the ferum from each, as free
from red blood as poffible, then put the red part of the one
into the ferum of the other, and fhook them to mix them
well ; and, upon letting them ftand quiet, the red globules
appeared to fall equally faft. From thefe experiments it
appears, that the red part of inflammatory blood was heavier than that which is not fo, and the ferum was lighter,
and the difference pretty nearly equal ; for if we could fuppofe that the red globules were one-tenth heavier, and the
ferum one-tenth lighter, then the difference in the fubfiding
of the red globules of inflammatory blood in its own ferum,
to that which is not inflammatory, would be as one to five;
and, if they were to be changed, then they would be equal.
To fee whether the blood from an inflamed part was different from that drawn from a part not inflamed, the following experiments were made :

A large leech was applied to an inflamed furface, and
when it had fucked itfelf full, another leech was fuffered
to fill itfelf from the breaft where no inflammation exifted;
they were both cut in two, and the blood received in two
tea-cups, kept moderately warm in a difh of warm water ;
both of them coagulated without the ferum feparating ; but
the inflamed blood was evidently of a lighter colour than
the blood from the uninflamed part ; but neither had the appearance of a buffy coat.

Whether the difpofition for inflammation, and the
change produced in the blood, arife from a real increafe of

animal life, or whether it is only an increafe of a difpofition
to act with the full powers which the machine is already
in poffeffion of, is not eafily determined ; but it appears to
be certain, that it is either the one or the other : there are
fome circumftances, however, that would incline us to fufpect it to be the latter, becaufe there is often inflammation
when the powers of the machine are but weak, where it appears to be only an exertion of very weak powers, arifing
from fome irritation produced ; in fuch cafes the blood will
fhew figns of weaknefs although fizy.

This appears to be equally the cafe in local inflammation,
and inflammatory fevers, or in the fymptomatic fever*.
That it is an increafe of the one or the other, and that the
fenfible effect produced arifes from the action taking place,
both in the folids and fluids, is proved by the method of
treatment, which will be further illuftrated in fpeaking of
the mode of cure: on the other hand, where there is great
debility in the folids, where the powers of prefervation (the
firft animal powers) are weak, therefore the action weak,
and where of courfe the body muft have a tendency to diffolution, there we find the very reverfe of the former appearance in the blood. Inftead of feparating diftinctly, and
coagulating firmly, we have the whole mafs of blood keeping
mixed, and hardly any coagulation, only becoming of a
thicker confiftence.

* On the other hand it would appear reafonable to fuppofe,
that there was really an increafe of animal life, for women who
are breeding, and are in perfect health, always have fizy blood ;
and this is moft remarkably the cafe with all animals in fimilar
fituations ; now it would appear neceffary for an animal, whenever put into a fituation where greater powers are wanted, to
have thefe powers increafed. In a breeding woman there is a
procefs going on, though natural yet uncommon, and which requires a greater exertion, or a greater quantity of powers than
ufual, and therefore we have them produced. This procefs of
breeding although in many of its fypmtoms it is fimilar to fever,
is yet very different ; for actual fever kept up for nine months
would deftroy the perfon, while on the other hand, many are
relieved by fuch a procefs.

If thefe obfervations are juft, this blood fhould not be called
inflammatory blood, but blood whofe powers of life are increafed.

F

This effect, or appearance often takes place in those who
die inftantaneoufly. I fufpect that in fuch cafes the blood
dies firft, and alfo inftantaneoufly.

In the commencement of moft difeafes, and even through
the whole courfe of many, the fituation of the blood appears
to be an object with nature. In fome the blood forfakes the
fkin and extremities, and we may fuppofe the fmaller veffels
in general; for when we can obferve internal parts, fo we find
it, fuch as the mouth in general, eyes, etc. a general palenefs
takes place, which is beft feen in the lips, and even a fhrink-
ing of the external vifible parts takes place, efpecially the
eyes, fo that the perfon looks ill, and often looks as if dying.
The pulfe is at this time fmall, which fhows that the whole
arterial fyftem is in action.

This appears to arife from debility, or the want of powers
in the conftitution to be acted on by fuch a difpofition at
the time, fo that the whole powers or materials of life
are called into the vital parts or citadel, and the outworks
are left to themfelves. Such is the cafe with fainting; the
cold fit of an ague; the cold fit or beginning of a fever;
rigors or beginnings of exacerbations; it is alfo the cafe with
the hectic.

In the commencement of difeafes it does not appear to
arife from real debility of conftitution, but the novelty of
the action, and of courfe a debility in that action, and in that
only; but in the hectic, where a real debility has taken
place, thofe appearances are owing to that caufe; however,
even in the hectic, this debility is affifted by the unnatural-
nefs of the action.

In the firft, where there are real powers, it would appear
as if nature was ftruggling with the new difpofition, and it
either becomes deftroyed entirely or in part, and the blood
is then determined to the fkin; and we may fuppofe into
the fmaller veffels in general; then the pulfe becomes full;
the whole action now appears to be there, and it becomes
hot; when that action in the fkin ceafes, a perfpiration
takes place, and nature feems in many cafes to be at reft;
in fome diforders this ceffation is perfect for a time, as in
agues; fometimes wholly, as in flight colds; but often
imperfectly, as in continued fevers, where the fenfation
appears only to arife from wearinefs, which prevents the
continuance of the action, not from an alteration of the
difpofition.

In other difeafes the blood is thrown very early upon the exterior parts. The face fhall look bloated, the eyes full, the fkin red, dry and hard to the touch.

Thefe fymptoms, I fufpect, belong more to fevers of the putrid kind, and have lefs connexion with furgery than the former.

The pulfe is often as ftrong a fign of the ftate of the conftitution as any other action that takes place in it, though it is not fo always; but as the pulfe has but one circumftance attending it, that we can really meafure, all the others being referable to the fenfation or feeling of the perfon who is to judge of it, the true ftate of the pulfe is not eafily afcertained. The knowledge of the foft, the hard, and the thrill, are fuch as can only be acquired with accuracy by the habit of feeling pulfes in thefe different ftates, and by many is not to be attained; for fimple fenfation in the minds of any two men are feldom alike. Thus, we find, it happens with refpect to mufic; for what would be difagreeable and not in harmony to one ear, which is nice and accuftomed to the harmony of founds, will not be fo to another.

The late Dr. Hunter was a ftriking inftance of this ; for though he was extremely accurate in moft things, he could never feel that nice diftinction in the pulfe that many others did, and was ready to fufpect more nicety of difcrimination that can really be found. Frequency of pulfation in a given time is meafurable by inftruments ; fmartnefs or quicknefs in the ftroke, with a paufe, is meafureable by the touch ; but the nicer peculiarities in the pulfe are only fenfations in the mind. I think I have been certain of the pulfe having a difagreeable jar in it when others did not perceive it, when they were only fenfible of its frequency and ftrength ; and it is perhaps this jar that is the fpecific diftinction between conftitutional difeafe or irritation and health ; frequency of pulfation may often arife from ftimulus, but the ftroke will then be foft ; yet foftnefs is not to be depended on as a mark of health, it is often a fign of diffolution ; but there muft be other attending fymptoms.

In the confideration of the peculiarities of the pulfe, it is always neceffary to obferve, that there are two powers acting to produce them, the heart and the arteries ; that

one part of the pulfe belongs to the heart alone, another
to the arteries alone, and the third is a compound of both;
but the actions of the heart and arteries do not always correfpond ; the heart may be in a ftate of irritation, and act
quickly in its fyftole, while the arteries may be acting flowly ; for the heart is to be confidered as a local part, while
the veffels muft be confidered as univerfal, or even conftitutional.   The ftroke, (which is the pulfe) with the number of them that are made in a given time, whence the pulfe
is commonly called quick or flow, their regularity and irregularity as to time, and the quicknefs of the ftroke itfelf,
belongs to the heart. The quicknefs of the heart's action
often takes place, although the pulfations are not frequent,
which gives a kind of reft or halt to the artery or pulfe, efpecially if the pulfe be not frequent.   The hardnefs, the
vibratory thrill, the flownefs of the fyftole, with the fullnefs and fmallnefs of the pulfe belong to the arteries.   As
the pulfe arifes from the actions of the folids or machine,
its ftate will be of courfe according to the nature of the machine at the time, and therefore is capable of being in one
of thefe ftates, natural and difeafed.

    In moft difeafes of the conftitution, whether originating
from it, or arifing in confequence of difeafes of parts,
where the conftitution becomes affected by fympathy, the
pulfe is altered from a natural to a difeafed ftate, the degree of which will be regulated by thofe affections.   This
a'.eration is commonly fo conftant, and fo regularly of the
nature of the difeafe, that it is one of the firft modes of
intelligence we have recourfe to, in our inquiries into its
nature ; but alone it is not always a certain guide ; for
where there are peculiarities of conftitution, we find the
pulfe correfponding to thofe peculiarities, and, perhaps, in
direct contradiction to the accuftomed ftate of the local affection.   The fame parts too, under difeafe, give very irregular, or uncertain figns in the actions of the heart and
veffels, fuch as difeafes, or injuries done to the brain.

    The varieties which the pulfe admits of are feveral.   It
is increafed in its number of ftrokes, or it is diminifhed.
It is regular or irregular, as to time, in its ftrokes ; it is
quick in its ftroke, or diaftole, and flow in its fyftole.   It
is hard in its diaftole, and it vibrates in its diaftole.

    In moft cafes, probably where the conftitution is in a
ftate of irritation, the pulfe will be quick and frequent in

its number of ftrokes in a given time, and the artery will become hard from a conftant, or fpafmodic contraction of its mufcular coats, fo as to give the feel of hardnefs to the touch ; befides which, the diaftole of the artery is not regularly uniform ahd fmooth, but preceeds by a vaft number of ftops, or interruptions, which are fo quick as to give the feel of a vibration, or what we would exprefs, by a thrill.

The pulfe under fuch a difpofition, or mode of action, may be either full, or fmall.

Thefe two very oppofite effects do not feem to arife from a difference in the quantity of blood, which might at firft be fuppofed; I fhould rather fufpect that they arife from a difference in the degrees of ftrength, which will be more or lefs, according to the nature of the parts inflamed, and the degree of irritability of the patient at the time. Thefe give, more or lefs, an anti-diaftolic difpofition to the arteries; and while the arteries have the power of contraction, and are in a ftate of irritation, this effect will always take place.

It is certain at leaft, that the arteries do not commonly, in fuch a ftate of conftitution, dilate fo freely and fo fully as at other times, and as this will vary very quickly, (if the conftitutional irritation varies quickly) it is more reafonable to fuppofe, that it is an immediate effect of the arteries, than an increafe, and decreafe of the quantity of blood.

If this be really the cafe, then we fhould naturally fuppofe that the motion of the blood in the arteries would be increafed in proportion to their diminifhed fize ; except we fhould alfo fuppofe that the diaftole, or the fyftole, or contraction of the heart, is alfo diminifhed in the fame proportion. The firft of thefe, I think, may probably be the cafe, as we find that the blood forfakes the furface of the body in fuch a ftate of the conftitution, as will hereafter be obferved, therefore muft be collected in the larger veins about the heart.

If the heart was to dilate and throw out its whole contents at each fyftole, then the velocity of the blood, in the arteries under fuch a ftate of contraction of arteries would be immenfe, and it might then be pufhed into the fmaller veffels on the furface of the body, which it certainly is not.

The quick, hard, and vibratory pulfe is generally an attendant upon inflammations ; and whether it be attended with fullnefs, or the contrary, depends a good deal upon the part that is inflamed, which either increafes, or decreafes the irritability, which will be defcribed in treating of the different parts inflamed.

In fuch a ftate of the conftitution, as produces fuch a pulfe, the blood, which appears to be only a paffive body, acted upon by the heart, fo as to produce the diaftole of the artery, and reacted on by the veffels, making the complete pulfe, this blood, I fay, is generally found in a different ftate from that where there are not thefe fymptoms in the pulfe ; they, as it were, conftantly attend each other, or are the reciprocal caufe and effects of one another, as was taken notice of when I was fpeaking of the ftate of the blood in inflammation.

From the account I then gave of the ftate of the blood in inflammation, and have now given of the pulfe, under the fame action, it fhould naturally be expected, that they fhould explain each other ; which, for the firft part they certainly do ; yet, thefe appearances of the blood, and the kind of pulfe, are every now and then appearing to be in oppofition to each other, in their common attending circumftances ; but this cannot be known till the perfon is bled ; when the pulfe is quick and hard, with a kind of vibration in the action, we generally have fizy blood. This may arife from fever, or fuch inflammation, etc. as affects the conftitution, or vital parts, thefe being fo difeafed as to keep up a conftitutional irritation, which will always be an attending fymptom ; but, when we have neither a quick, nor hard pulfe, both, perhaps below par, and rather fmall, no vifible fever, nor inflammation, but, probably, fome ftrong, undetermined fymptoms, fuch as pain, which is moveable, being fometimes in one place, fometimes in another, but at the fame time feeming to impede no natural function, yet upon bleeding, the blood fhall be fizy, and the fize fhall have ftrong powers of contraction fo as to cup.

A gentleman was ill with a pain, chiefly in his right fide, but upon the part being rubbed, or application being made to it, the pain feemed to move to fome other part ; from which circumftance it was fuppofed to have connexion with the bowels ; at other times he was tolerably well. His pulfe was flow, fmall, and foft, and not at all, to the feel, like a pulfe which required bleeding. He defired to be bled, and when bled, the blood was extremely fizy ; the fize being ftrong, and contracting fo much as to draw in the edges, forming the upper furface into a hollow, or cup. His pulfe became fuller, quicker, and harder ; he was bled a fecond time; the blood was the fame, and the above fymptoms increafed fo much; that I obferved, immediately after the fecond bleeding, his pulfe was quicker, harder, and fuller, than it was juft before the bleeding. That it might be quicker and fuller, I could conceive, becaufe I have often feen fuch effect from bleeding, where there had been an oppreffed and languid pulfe ; but I cannot fay that I ever faw a cafe where the pulfe became harder, and acquired the vibration, except when debility, or languor was produced, and where the blood was weak in its powers of coagulation, being flat on the coagulated furface. Another want of correfpondence, or irregularity, takes place when a conftitution fympathizes with a local inflammation. There are cafes where the pulfe becomes flow, and often irregular ; fuch are moftly to be found in all people, when the conftitution is affected either originally or fympathetically, and in fuch, I fufpect that a difpofition for diffolution, and perhaps mortification, is much to be feared.

A man, aged fixty-eight years, had an occafional inflammation in one of his legs, which often ulcerated, and which feemed to arife more from a defect in the conftitution than to be fimply local. In thofe indifpofitions, his pulfe feldom exceeded forty in the minute, and as he began to get better, his pulfe became more and more frequent.

The varieties of the pulfe arifing from the feat of the inflammation, and the nature of the part inflamed, will be expreffed when I treat on inflammations in different fituations and parts.

VIII. THE EFFECTS OF INFLAMMATION ON THE CONSTITUTION, ACCORDING TO THE STRUCTURE OF PARTS, SITUATION OF SIMILAR STRUCTURES, AND WHETHER VITAL OR NOT VITAL.

THESE circumftances make a very material difference in the effects on the conftitution, arifing from local inflammation ; for we fhall find that the effects on the conftitution are not fimply as the quantity of inflammation, but according to the quantity and parts combined, (fuppofing conftitutions to be equal) which I fhall now confider feparately.

In common parts, as mufcle, cellular membrane, fkin, etc. the fymptoms will be acute; the pulfe ftrong and full, and the more fo, if it be felt near to the heart; but perhaps, not fo quick as when the part is far from it ; fince there will be lefs irritability.   The ftomach will fympathize lefs, and the blood will be pufhed further into the fmaller veffels.

If the inflammation is in tendinous, ligamentous, or bony parts, the fymptoms will be lefs accute, the ftomach will fympathize more, the pulfe will not be fo full, but perhaps quicker, becaufe there will be more irritability, and the blood will not be fo much pufhed into the fmaller veffels, and therefore forfake the fkin more.

It feems to be a material circumftance, whether the inflammation is in the lower or upper extremity ; that is, far from, or near to the heart ; for the fymptoms are the more violent, the conftitution is more affected, and the power of refolution feems to be lefs, when the part inflamed is far from the fource of the circulation, than when near it, even when parts are fimilar, both in texture and ufe.

Whatever courfe the inflammation is to run, or in what ever way it is to terminate, it is done with more eafe when near to the heart than when far off.

All the parts that may in one fenfe be called vital, do not produce the fame effects upon the conftitution ; and the difference feems to arife from the difference in their con-

nexions with the ftomach. It is to be obferved, that vital
parts may be divided into two, one of which is in itfelf immediately connected with life, as the ftomach ; the other,
where life only depends upon it, in its action or ufe ; the
heart, lungs, and brain are only to be confidered in this laft
light ; therefore, they have a confiderable fympathizing affection with the ftomach ; the fymptoms are rather depreffing ; the pulfe is quick, fmall, and the blood is not pufhed
into the fmaller veffels.

If the heart or lungs are inflamed, either immediately, or
affected fecondarily, as by fympathy, the difeafe has more
violent effects upon the conftitution than the fame quantity of inflammation would have, if it was not in a vital
part, or was in one with which the vital parts did not fympathize ; for if it is fuch as the vital parts fympathize with
readily, then the fympathetic action of the vital parts will
affect the conftitution, as in an inflammation of the tefticle.

The pulfe, in fuch cafes, is much quicker and fmaller than
when in a common part, as a mufcle, cellular membrane, or
fkin ; but not fo much fo as in the ftomach, and the blood
is more fizy.   When the inflammation is in the heart only,
its actions are extremely agitated and irregular.   If in the
lungs, fingly, the heart in fuch cafes would appear to fympathize, and not allow of a full or free diaftole.

The ftomach does not in common fympathize in fuch cafes, which is the reafon, perhaps, of the inflammation not
depreffing ; but it is to be obferved, that I make a material
difference between the inflammation of the lungs, commonly called a pleurify, and thofe difeafes that begin flowly, and fpin out to great lengths, and which are truly fcro-
phulous, producing the hectic ; for in them we have the
hectic pulfe, and not the inflammatory.

If the ftomach is inflamed, the patient feels an oppreffion
and dejection through all the ftages of the inflammation ;
fimple animal life feems to be hurt and leffened, juft as fenfation is leffened when the brain is injured ; the pulfe is
generally low and quick, the pain is obtufe, ftrong, and op-
preffing, fuch as a patient can hardly bear.

If the inteftines are much affected, the fame fymptoms
take place, efpecially if the inflammation be in the upper
part of the canal ; but if it is the colon only which is af-

fected, the patient is more roufed ; and the pulfe is fuller
than when the ftomach only is inflamed.

If it be the uterus, the pulfe is extremely quick and low.
If it be a tefticle that is inflamed, the pain is depreffing,
the pulfe is quick, but not ftrong.

When the inflammation is either in the inteftines, tefticle, or uterus, the ftomach generally fympathizes with
them, which will produce, or increafe the fymptoms peculiar to the ftomach.   In inflammation of the brain, I believe the pulfe varies more than in inflammations of any
other part ; and, perhaps, we are led to judge of inflammation there, more from other fymptoms than the pulfe.   I
believe the pulfe is fometimes quick, flow, depreffed, full,
etc. and which may accord with the other fymptoms, fuch
as delirium, ftupor, etc.

It is to be obferved, when the attack upon thefe organs,
which are principally connected with life, proves fatal, that
the effects of the inflammation upon the conftitution run
through all the ftages with more rapidity than when it happens in other parts ; fo that at its very beginning, it has
the fame effect upon the conftitution, which is only produced by the fecond ftage of fatal inflammation in other parts.
Debility begins very early, becaufe the inflammation itfelf is
interfering immediately with the actions of life ; and alfo
in fuch parts univerfal fympathy takes place more readily,
becaufe the connexion of thefe parts by fympathy is more
immediate ; and if the fympathy is fimilar to the action,
then the whole is, in fome degree, in the fame action.

If the inflammation comes on in a part not very effential
to life, and with fuch violence as to affect the actions of
life, or to produce univerfal fympathy, the pulfe is fuller
and ftronger than common ; the blood is pufhed further into the extreme arteries than when the inflammation is in a
vital part ; the patient, after many occafional rigors, is at
firft rather roufed, becaufe the actions of the part are
roufed ; and the effects in the conftitution are fuch
as do not impede any of the operations of the vital
parts.   It is allowed to proceed to greater lengths, or
greater violence in itfelf, before the conftitution becomes
equally hurt by it ; and the conftitutional fymptoms produced at laft, may be faid to arife fimply from the violence
of the inflammation ; but this will take place, more or
lefs, according to circumftances ; it will be according to

the nature of the parts, whether active as mufcles, or inactive as tendons ; alfo according to the fituation of the fame
kind of parts, as well as according to the nature of the conftitution. If the conftitution is ftrong, and not irritable,
the pulfe will be as above ; but if the conftitution is extremely irritable and weak, as in many women who live fe-
dentary lives, the pulfe may be quick, hard, and fmall, at
the commencement of the inflammation, fimilar to the inflammation of vital parts. The blood may be fizy, but
will be loofe and flat on the furface.

IX. GENERAL REFLECTIONS ON THE RESO-LUTION OF INFLAMMATIOM.

I now come to the moft difficult part of the fubject, for
it is much more eafy to defcribe actions, than to affign motives ; and without being able to affign motives, it is impoffible to know when or how we may or fhould check actions
or remove them. I have endeavoured to fhew, that an
animal body is fufceptible of impreffion, producing action :
that the action, in quantity, is in the compound ratio of
the impreffion, the fufceptibility of the part, and the powers of action of the part or whole ; and in quality, that it
is according to the nature of the impreffing power and the
parts affected. I have alfo endeavoured to fhew that impreffions are capable of producing, or increafing natural actions, and are then called ftimuli : but that they are likewife
capable of producing too much action, as well as depraved,
unnatural, or what are commonly called difeafed actions.
The firft of thefe I have mentioned by the general term, irritations : the depraved, etc. come in more properly in treating of peculiar, or fpecific actions.

Since then an animal body can be made to increafe its
natural action, or to act improperly by impreffion, fo we
can fee no reafon, when it is acting too violently, why it
fhould not be reftrained by impreffion ; or when acting improperly, in confequence of thefe impreffions, why it fhould
not be made to act properly again by the fame mode, namely, by impreffions.

These modes of action we are first to understand, and then the power of correcting, or counteracting those impressions, in order to diminish, or prevent the action, so as to produce one that is healthy or natural; besides, an injury which produces a new mode of action, and a disease, which is a new mode of action, often happen when the machine is in perfect health, and in such a state as is perfectly in harmony with that health; but which state is not suitable to disease; therefore, it is to be presumed, the more perfect health the body enjoys, the less it bears a change in its actions. Thus we know, that strong health does not bear considerable injuries, such as accidents, operations, etc. A man in strong health, for instance, will not bear a compound fracture in the leg, or an amputation of the same, so well as a man accustomed to such diseases, and reduced by them. We find, commonly, that our artificial mode of reduction, is by far too quick, and is almost as much a violence on the constitution, as the injury; when, therefore, considerable injuries or diseases commence, the constitution is to be brought to that state which accords best with that accident, or disease. The knowledge of that state of the body, at that time, as well as the operations of the whole animal, or of its parts, when arising from a disturbed or deranged state, or a diseased disposition, are to be considered as the first step towards a rational cure: but this alone is insufficient; the means of bringing the body to that state are also necessary which will include the knowledge of certain causes and effects, acquired by experience, including the application of many substances, called medicines, which have the power of counteracting the action of disease: or of substances perfectly inefficient in themselves, but capable, under certain circumstances, of producing considerable effects such as water, when hot and cold; or a substance when it varies its form, as from fluidity to vapour. Of these virtues we know nothing definitely; all we know is, that some are capable of altering the mode of action, others stimulating, many counter-stimulating: some even irritating, and others quieting, so as to produce either a healthy disposition and action in a diseased part, or to change the disease to that action which accords with the medicine, or to quiet where there is too much action; and our reasoning goes no further than to make a proper application of those substances, with these virtues. The difficulty is to ascertain the

connexion of fubftance and virtue, and to apply this in re-
ftraining or altering any difeafed action ; and as that cannot be demonftrated a priori, it reduces the practice of
medicine to experiment ; and this not built upon well determined data, but upon experience, refulting from probable data. This is not equally the cafe through the whole
practice of medicine, for in many difeafes we are much
more certain of a cure than in others ; but ftill, even in them
the certainty does not arife from reafoning upon any
more fixed data, than in others, where the certainty of a
cure is lefs ; but it arifes from a greater experience alone ;
it is ftill no more than inferring that in what is now to
be tried, there is a probable effect or good to arife in the
experiment, from what has been found ferviceable in fimilar cafes : difeafes, however, of the fame fpecific nature,
not only vary in their vifible fymptoms or actions, but in
many of thofe that are invifible, arifing probably from peculiarities of conftitution, and caufes, which will make
the effects of application vary, probably, almoft in
the fame proportion ; and as thofe varieties may
not be known, fo as either to adapt the fpecific medicine to them, or to fuit the difeafe to the medicine, it will then be only given upon a general principle,
which of courfe may not correfpond to the peculiarities.
Even in well-marked fpecific difeafes, where there is a fpecific
remedy, we find that there are often peculiarities, which
counteract the fimple fpecific medicine. This we even
fee in poifons, the moft fimple fpecific of all, becaufe its effect arifes in all cafes from one caufe ; the peculiarities,
therefore, in the difeafe muft arife from a peculiarity in
the conftitution, and not from the caufe of the difeafe.

The inflammation I have been treating of, is the moft
fimple of any, becaufe it is the fimple action of the parts
unmixed with any fpecific quality, arifing from caufes of
no fpecific kind, and attacking conftitutions, and parts, not
neceffarily having any fpecific tendency ; the cure, therefore,
or method of terminating the inflammation, which is called
refolution, (in cafes that will admit of it) muft alfo be very
fimple, if we knew it ; and accordingly, when the cure of
fuch is known, it lays the foundation of the general plan for
the treatment of all inflammations of the fame kind : but it
very rarely happens that a conftitution is perfectly free

* Vide the varieties of the inflammation, in the introduction.

from a tendency to some disease*; we seldom, therefore, see simple salutary actions of parts tending to relieve themselves from a violence committed : some constitutions being so irritable, that the inflammation has no disposition to terminate, and others so indolent that the inflammation passes into another species, as into scrophula ; all of which will require very different treatment.

The same varieties take place in specific inflammations; as also in inflammations arising from poisons ; for many will have the true inflammatory disposition joined with the specific ; in such, therefore, the same plan is to be pursued with the addition only of the specific treatment; but this must not be omitted, as the inflammation depends upon the specific disease. It is this critical knowledge, which becomes the basis of practice; and it is this which requires the greatest sagacity; and I must own, it requires more knowledge than comes to the share of most practitioners. As every inflammation has a cause, that cause should be removed before resolution can take place ; for the animal œconomy having a disposition within itself to discontinue diseased action, that of course subsides upon the removal of the cause ; and this disposition is so strong in some, as to appear to act alone. That removing the cause is a mode of resolution, is proved in the venereal bubo ; for by taking off the venereal action with mercury, the inflammation subsides, if another mode of action does not arise*. Inflammation, where it must suppurate, is most probably a restorative act, and cannot be resolved in those cases where restoration becomes necessary ; as for instance, in a wound that is kept exposed, the inflammatory act of restoration becomes, or is rendered, necessary, and it takes place ; but bring those parts together, or let the blood coagulate and dry upon it, and it becomes unnecessary. I have already observed, when treating of the causes of inflammation, which might be called the spontaneous, that they probably arose from a state of parts, in which they could not exist, similar to exposed surfaces, and therefore this act of restoration became necessary : if this be true, then probably by altering that state of parts, as we can by bringing the divided parts together, the inflammation would either not rise, or immediately cease ; but as we are not in all cases acquainted with the mode of restoring those

* Vide Treatise on Venereal Disease.

natural actions, we are obliged to be reftricted to thofe methods that render them eafier under this ftate, and which are often capable of turning the balance in favor of refolution.

As inflamed parts are not always vifible, it becomes neceffary that we would have fome rule to inform us, whether the part is inflamed or not; to afcertain which, we muft have recourfe to all the fymptoms formely mentioned, except the vifible ones. We ought alfo to have a guide refpecting the kind of inflammation, more efpecially as it is not fufficient, in many cafes, to be guided entirely by the appearances, even where it is in fight; it is often, therefore, very neceffary to inquire into the caufe of the inflammation, the nature of the conftitution, the effects that former inflammation has produced, and even into the temper and mind of the patient.

The cure of inflammation is refolution; and the attempt towards it, is principally to be made when the inflammation is in the adhefive ftate; for we find that often it goes no further, but fubfides, and this is refolution; probably the • fooner after its commencement it is the better. The ob- , ject of the attempt is to prevent fuppuration taking place, although fuppuration may be confidered as a refolution, but it is the mode of refolution we commonly wifh to avoid. Refolution is in general only to be attempted, with any probability of fuccefs, under the following circumftances: when the inflammation is in confequence of the conftitution, or a difeafe of the part: fecondly, in cafes of accident, where there is neither no expofure, or where it has been removed in time: as, for example, by bringing the parts in contact: thirdly, where the life of the part has not been deftroyed. In all fuch cafes we find that refolution can take place; but in thofe cafes arifing from accident, and a continuance of expofure joined, or where death of the parts is produced by the accident, it becomes impoffible to hinder the fuppuration from taking place.

I have already obferved, that in many bruifes, as well as fimple fractures, where the cavities are not expofed, and where they are to heal by the firft and fecond intention, the inflammation, in moft of thefe cafes, is capable of being refolved; although, in fome fuch cafes, the inflammation runs fo high as to threaten fuppuration. I have alfo already fhewn, that in parts which have been divided and expofed, the inflammation is, by bringing them together,

in a great meafure prevented ; or if it has taken place previous to the union, that the fame operation of union is
fufficient to produce refolution ; and I have likewife fhewn
that where parts were not brought together, nature attempted to prevent inflammation, by covering the wound
with blood, and forming an efchar, which, in many cafes,
will either prevent, or remove inflammation ; all of which
fhews a power of refolution, even in the cafes where the
parts have been expofed.

As it is commonly fuppofed that there are a great many
local difeafes that fhould not be refolved, the firft thing neceffary to be confidered, is, when the refolution fhould or
fhould not be attempted.   On the contrary, there are
cafes where inflammation is to be excited, but thefe arife
commonly from difeafe, which is not to our prefent purpofe ; yet it fometimes happens in accidents, where inflammation is neceffary, that it is not fufficient for the re-
inftatement of the injured parts, as in fome fimple fractures, where the firft bond of union, the extravafated blood,
had not fulfilled its purpofe, and had been abforbed, and
where the inflammation was too flight to fupply its place ;
fo that union of parts was prevented, and another mode
became neceffary, not at all à confequence of inflammation, viz. granulations without fuppuration ; all of which
retards ftill more the reftoration of the parts.   As this defect can only be known in bone, and in the foft union of
the bone, which is fimilar to the union in the foft parts, it
is reafonable to fuppofe, it may alfo take place in the
foft parts ; more efpecially thofe which are tendinous, or
ligamentous, where we find recovery very flow, for the
foft union in bones differs in nothing from that of the foft
parts ; it may, therefore, be a much more common defect
than is generally imagined.   In fuch cafes, if it could be
known, it would be proper to encourage, or even excite
inflammation.   If we cannot, probably, in any cafe determine where it fhould be excited, nor even where it fhould
be checked, yet we can fay in many cafes, where it is
unneceffary to check it.   Before we attempt to check inflammation, we fhould have reafon to fuppofe it is going further than is neceffary for the natural cure ; and
therefore it is laying the foundation of work for the furgeon.   It may be very difficult to fay, in many cafes,
when it fhould be checked.   The moft fimple reafon

will be to leffen pain, arifing in a part not merely when moved or touched, but in the act of inflammation. Secondly, where it may be uniting parts, the union of which we wifh to avoid; but this is an uncertain guide, even if we knew adhefions were taking place, for adhefions often prevent fuppuration. Thirdly, to prevent the inflammation from fuppurating; and in this laft, although the moft obvious, yet there is lefs certainty how far we may advife the attempt. It is alfo the moft difficult to effect; for in many cafes of fpontaneous inflammation, if it arifes from a ftate under which the parts cannot exift, nor their functions go on, fimilar to an expofed breach in the folids, then refolution fhould not be attempted: it may be palliated when going beyond what is neceffary for fuppuration; but when this practice is carried farther, it rather retards that falutary procefs. From the foregoing ftatement of particulars, it muft appear, that in many cafes it is unneceffary to check inflammation; in others it would be wrong, and in many very neceffary and probably the beft guide is its going further than appears from the caufe to be falutary; yet in practice we find applications, and other modes of refolution, immediately had recourfe to, which muft be confidered as opprobrious to furgery.

Inflammations, in confequence of accidents, ought in general to be refolved, if poffible. It is perhaps impoffible to produce a fingle inftance where a contrary practice would be preferable, except as above related, where its confequence would be to anfwer fome great purpofe; and it is alfo conceivable, that this local difeafe, produced by accident, might relieve the conftitution from fome prior diforders, fimilar to what is underftood to be the effects of an iffue. Mr. Foote was relieved of head-achs, of long ftanding, by the lofs of a leg, which may be confidered as a proof of this; but he afterwards died of a complaint in his head, very fimilar to an apoplexy. It might be fuppofed, on the other hand, that the temporary cure was the caufe of the apoplexy.

Inflammation, in confequence only of a difeafe in a part, appears to be under the fame circumftances, with refpect to refolution; but an inflammation arifing from a preceeding indifpofition in the conftitution (commonly called critical) has always been claffed among thofe which fhould not be cured locally, and this has got the term of repulfion : it has been infifted on, that the inflammation fhould rather be encouraged, and fuppuration produced, if poffible. If the inflammation is really a concentration of the conftitutional complaint, and that by not allowing it to reft here, the fame difpofition is really diffufed over the whole animal again, and at liberty to fix on fome other part, it certainly would be better to encourage its ftay ; but, in fuch cafes it is always underftood that the inflammation is in fuch parts as will readily admit of a cure when fuppuration takes place ; for if the difeafe be otherwife fituated, then the cure of the conftitution by fuppuration will be a mode of cure which will reflect back another difeafe upon it, under which it will fink : refolution of inflammation, therefore, in the firft of thefe fituations fhould, if poffible, be brought about. For inftance, many deep-feated inflammations, if allowed to fuppurate, would of themfelves moft certainly kill. This might be illuftrated by the gout, when either in the head or ftomach, for when in fuch parts it had better be repelled, and left to find another part lefs connected with life; which, if in the feet, would be called repelling of it ; but ftill it does not appear to me neceffary that it fhould fuppurate, for fuppuration is only a confequence of the inflammation, and not an immediate confequence of the original or conftitutional difeafe, but a fecondary one* : as fuppuration, therefore, is only a thing fuperadded, and as we fhall find that inflammation generally fubfides when fuppuration comes on, I fee no reafon why inflammation in the prefent cafe, fhould not as well fubfide by refolution as by fuppuration : however, it may be fuppofed, that although fuppuration is not the natural, or immediate effect of the difeafe, yet as it is a continued local

---

* This is contrary to the common received opinion, but it is according to my idea of fuppuration, for I have all along confidered inflammation as the difeafe, and fuppuration only as a confequence of that difeafe ; and have fuppofed the difeafe to be gone when fuppuration has taken place ; but, according to the common opinion, fuppuration was the thing to be wifhed for ; becaufe all difeafes arofe from humors ; but as we have not once mentioned humors, and therefore made it no part of our fyftem, we muft alfo drop it at prefent.

action, and the thing fought for by the conftitution, and as inflammation muft precede it, the parts muft fubmit to thofe regular proceffes; for it muft be fuppofed to be capable of diverting the difeafe to this part.

X. OF THE METHODS OF RESOLUTION BY CONSTITUTIONAL MEANS.

THE firft thing to be confidered is the kind of inflammation; when vifible, which will in fome degree fhew the kind of conftitution; the next is the nature of the part inflamed, and the ftage of inflammation; for upon thefe depend in fome meafure the method of relief. In cafes of ,expofed internal furfaces, the inflammation cannot be refolved, becaufe the caufe ftill exifts till inflammation has refolved itfelf; but it may be leffened, and this probably takes place by leffening every thing which has a tendency to keep it up; and in all likelihood, little more can be done in fpontaneous inflammations; for as yet we know of no method which will entirely quiet or remove the inflammatory difpofition or mode of action, as there is no inflammatory fpecific with which we are acquainted. When I defcribed inflammation, I obferved there was either an increafe of life, or an increafed difpofition to ufe with more violence the life which the machine, or the part was in poffeffion of; and alfo there was an increafed fize of veffels, and of courfe an increafed circulation in the part inflamed, and in the conftitution in general. If this theory of the mode of action of the veffels in inflammation is juft, then our practice is reducible to two principles; one confifting in removing the caufe of that action; the other in counteracting the effect. In the firft, as we feldom know the caufe, but only fee the effect, except in fome fpecific difeafes, for which we have a fpecific remedy, we do not know with any degree of certainty how to act; but as the fecond, that is, the effect, is more an object of our fenfes, we can apply with more certainty our reafoning upon it; for reafoning from analogy will affift us in our attempts. We find, from common obſervation, that many circumſtances in life, as alſo many applications to parts, will call forth the contraction of the veſſels : we are, from the above theory, to apply ſuch means ; and whatever will do this, without irritation, will ſo far counteract the effect*. I have already obſerved, that wherever there has been a violence committed, or ſome violent action is going on, there is a greater influx of blood to that part. Leſſening, therefore, that influx becomes one mode of relief ; for as the veſſels dilate, they ſhould not be encouraged in that action. Although the increaſed influx is to be conſidered chiefly as an effect, yet it is to be conſidered as a ſecondary cauſe ; and from our ignorance of the immediate cauſe, it is probably only through ſuch ſecondary cauſes that we can produce any effect ; and upon theſe principles, moſt likely reſt, in ſome meaſure, the method of reſolution ; for whatever will leſſen the power and diſpoſition, will alſo leſſen the effect ; and poſſibly theſe will likewiſe leſſen the force of the circulation.

If the inflammation is attended with conſiderable action and power, as it were, increaſing itſelf, then the modes of reſolution are to be put in practice ; the one by producing a contraction of the veſſels, the other by ſoothing or leſſening irritability, or the action of dilatation.

The firſt, or contraction of the veſſels, is produced in two ways ; one by producing weakneſs ; for weakneſs excites the action of contraction of the veſſels ; the other, by ſuch applications as induce the veſſels to contract.

The means of producing abſolute weakneſs are ! bleeding and purging ; but the bleeding alſo produces irritability for a time, and is often attended by a temporary weakneſs of another kind, viz. ſickneſs.

The inconvenience, however, ariſing from this practice is, that the ſound parts muſt nearly, in the ſame proportion ſuffer with the inflamed ; for, by bringing the inflamed part upon a par with health, the ſound parts muſt be brought much lower, ſo as to be too low. The ſoothing may be

---

* As this is a new theory of the action of the veſſels in inflammation, and the only one that can poſſibly direct to a method of cure, it is to be hoped that attention will be paid to it ; and, if juſt, that more certain methods of reſolution will be diſcovered.

producing by fedatives, relaxants, antiftimulents, etc. fuch as many fudorifics, anodines, etc.

The firft method will have the greateft, the moft perma-nent, and the moft lafting effect ; becaufe, if it has any ef-fect at all, the difenfed action cannot be foon renewed. The fecond will act as an auxiliary ; for fo far as irritation is a caufe, this will alfo leffen it ; and the two fhould go hand in hand ; for wherever we leffen power, we fhould, at the fame time, leffen the difpofition for action, or elfe we may increafe the difpofition ; but neither bleeding, purging, nor ficknefs, can poffibly leffen the original inflammatory difpo-fition ; for none of them will refolve a venereal inflamma-tion, when mercury will: nor will they refolve the eryfepe-latous inflammation, although that inflammation has the very action for which we fhould bleed in the common in-flammation, viz. dilatation of veffels. However, thefe means may, in fome fenfe, be reckoned direct ; for whate-ver will produce the action of contraction in the veffels is counteracting the action of dilatation. Leffening the power of action belonging to any difpofition, can only leffen or pro-tract the effects, which, however, will be of fingular fervice, as lefs mifchief will be done, and it will often give the difpofition time to wear itfelf out. Means employed, on this princi-ple, fhould be fuch as give the feel of weaknefs to the conftitution ; which will effect the part, and will make the veffels contract; but this practice fhould not be carried fo far as to produce the fenfe of too much weaknefs, for then the heart acts with great force, and the arteries dilate.

Bleeding then, as a general principle, is to be put in prac-ice ; but this muft be done with judgement; for I conceive the effects of bleeding to be very extenfive. Befides the lofs of any quantity of blood being univerfally felt, in propor-tion to the quantity loft, an univerfal alarm is excited, and a greater contraction of the veffels enfues, than fimply in proportion to this quantity, in confequence as it would ap-pear, of a fympathetic affection with the part bleeding.

Too much blood, in an inflammation, is a load upon the actions of the circulation. Too little, produces debility and irritability ; becaufe, there is a lofs of powers, with an increafed action to keep up, which is now not fupported. It would feem that violent actions of a ftrong arterial fyf-tem, required lefs blood than even the natural actions ; and even lefs ftill than a weak or irritable fyftem ; from whence we muſt ſee, that bleeding can either relieve inflammatory action, or increaſe it, and therefore not to be uſed at random.

As many patients that ſeem to require bleeding have been already bled, it may not be improper to inquire how they bear, or are affected by bleeding; for, certainly, all conſtitutions (independent of every other circumſtance) do not bear this evacuation equally, and it is probable, that its effects on inflammation may be nearly in the ſame proportion; if ſo, it becomes a very uſeful caution; for although the loſs of blood may, as a general principle, be ſet down as a weakener, and probably the greateſt, as we can kill by ſuch means, yet the loſs of certain quantities in many conſtitutions is neceſſary for health : this is either when there is a diſpoſition to make too much blood, or a conſtitution that cannot bear the uſual quantity; in ſuch, when known, bleeding with freedom is certainly neceſſary. If the inflammation is known to be attended with real powers, bleeding is abſolutely neceſſary, in ſuch quantity as to take off from the force of the circulation, which ariſes from too much blood; or if that is not ſufficient, then as much as will cauſe a contraction of the veſſels; but in caſes of too great an action of weak parts, then the proper quantity to be taken, is no more than may aſſiſt the dilatation of the veſſels, which will leſſen the violence of motion in the blood, or remove the ſenſation in the part inflamed of having too much to do; the quantity, therefore, muſt be regulated according to the ſymptoms, and other circumſtances; for inſtance, according to the viſible indications.

We are to remark here, that every part of the body, under inflammation, will not bear bleeding alike. I believe that the conſtitution bears bleeding beſt, when the inflammation is in parts not vital, and thoſe near the ſource of the circulation : whatever diſturbs ſome of the vital parts, depreſſes, but not equally in all; and in them it becomes more neceſſary to be particular, for in accidents of the brain, bleeding freely, even ſo as to produce ſickneſs and fainting, is neceſſary. It is probable, that the ſickneſs attending ſuch accidents, is deſigned to leſſen the influx to the head, and occaſion the veſſels of the brain to contract.

The indications for bleeding are firſt, according to the violence of the inflammation, joined with the ſtrength of the conſtitution, which will in general point out the kind of inflammation. Secondly, according to the difpofition to form much blood : thirdly according to the nature of the part, whether vital or not : fourthly, according to its fituation, in point of diftance from the heart : fifthly, according to the effect of the inflammation on the conftitution.

With regard to this evacuation, it is worthy of particular confideration, whether or not in all cafes, where it can be put in practice, bleeding in or near the part will anfwer better than taking the blood from the general habit ; for certainly lefs may be removed in this way, fo as to have equal effect upon the part inflamed, (and probably upon every other difeafe that is relieved by bleeding) and yet affect the conftitution lefs ; for although, in many cafes, the general habit may be relieved by bleeding, yet the part affected, where it can act, will in all cafes require this evacuation moft, and local bleedings will keep nearer thefe proportions, whereas taking blood from the general fyftem is juft the reverfe. That local bleeding has very confiderable effects on the inflamed part, is proved by the gout ; for applying leeches to the part inflamed, commonly relieves that part, and often almoft immediately*. We find that bleeding by leeches alone will remove a tumour in the breaft, having all the appearances of a fchirrus, which cannot be confidered as inflammatory ; its powers, therefore, extend beyond inflammation. We find relief by bleeding in the temporal artery, or jugular vein, for complaints in the brain ; or cupping and bleeding with leeches, on or near the part; as applying leeches to the temples in inflammations of the eye.

I have obferved that there is fomething fimilar to fympathetic affection in bleeding. I conceive that all the fympathetic powers, the univerfal, continued, and contiguous, may be brought into action from the local influence of bleeding. Thus, bleeding in the part inflamed, I can conceive, does more than fimply emptying the veffels mechanically, for that would be foon reftored from the general circulation ; but it acts by continued fympathy, viz. the veffels of the part being opened, they contract for their own defence, and this is carried further among the veffels of the part ; fo that bleeding from the part acts in two ways, viz. mechanically, by relieving the veffels of fome blood, fo

* It is not meant here to recommend bleeding in this difeafe

as allow them to contract in proportion as the load is taken off; and also to excite them to contraction, in order to prevent the effusion of blood. I suppose, likewise, that contiguous sympathy comes into action; for this would appear from practice and observation to be a principle in bleeding; therefore, in inflammation of contiguous parts, it is proper to bleed from the skin opposite to them, as from the skin of the abdomen, in complaints of the liver, stomach, and bowels; and likewise from the loins in inflammatory affections of the kidneys. In affections of the lungs, bleeding opposite to them is of service; but in such cases, it is not clear where the inflammation is; for if in the pleura, then it does not act upon this principle, but by continued sympathy: bleeding on the scalp relieves head-achs; and the relief given to the testicle by bleeding from the scrotum, inflammation of that body, proves the principle.

Where the first indication for bleeding takes place, viz. where there is violent inflammation, with strength of constitution, bleeding freely, will be of singular service. The same mode of practice is also to be followed under the circumstance of strength, with respect to the second, third, fourth, and fifth; but each will not require the same quantity to be taken under equal strength of constitution, as will be taken notice of when treating of them separately. As it seldom happens that bleeding once will be sufficient in a considerable inflammation, the first, or preceeding blood taken, becomes a symptom of the disease. If the coagulating lymph is long in coagulating, so that the globules have time to subside, there will be what is called a thick buff; and if its surface is considerably cupped, then future bleedings may be used with less caution; because such appearance indicates strong powers of coagulation, which always shews strength in the solids; but if the blood is weak in its powers of coagulation, lies flat in the dish, then we must be cautious in our future bleedings; or if it was strong at first in its powers of coagulation, and after repeated bleedings becomes weak, then we must not pursue this further; but in some cases it is proper to pursue it to this point, for we shall sometimes find that the inflammatory symptoms shall not cease after repeated bleedings, if the strength continues; but the moment a degree of looseness is produced in the blood, that moment will the inflammatory action cease. The following case is a strong instance of this. A lady had a violent cough, tightnefs in refpiration, lofs of appetite, ftron₂ fizy blood, and the fymptoms continued to the fixth bleeding, when the blood was not quite fo fizy; but the moft remarkable change was, its remaining flat on the furface. Upon this bleeding, all the fymptoms difappeared; and here, although the blood became weak in its power of coagulation, yet it did not produce irritability in the conftitution, the veffels of the inflamed parts having ftill had power to contract. On the other hand, there may be indications for bleeding fparingly : firft, when there is too much action, with weakened powers : fecondly, when there is a difpofition to form but little blood : thirdly, when the part affected is far from the fource of the circulation.

From the above three difpofitions that require bleeding fparingly, or with caution, I may obferve, that it will moft probably be proper in all fuch cafes to bleed from, or as near the part affected as poffible, in order to have the greateft effect, with the lofs of the leaft quantity of blood; more fo than when the conftitution is ftrong; becaufe the conftitution in fuch cafes fhould feel the lofs of blood as little as poffible; if from the part, leeches will anfwer beft, becaufe commonly little irritation follows the wound of a leecht : however, this can only be put in practice in inflammations not very remote from the furface. But in many cafes the blood cannot be taken away from the partitfelf, but only from fome neighbouring part, fo as to affect the part inflamed: thus, we bleed in the temporal artery for inflammation of the eyes; we bleed in the jugular veins for inflammation of the brain; and alfo in the temporal artery, to leffen the column of blood going to the brain, by the internal caro-

‡ However, this is not always the cafe; for it fometimes happens that an unkindly inflammation attends the wound, though not extenfive. It fometimes, alfo, happens that the lymphatic glands fwell in confequence of their bite; but thefe fo rarely occur, and are of fuch little confequence when they do that they are not to be regarded. From thence it has been conceived, that there was fomething poifonous in the bite of a leech; but I think there are no proofs of it : however, from another effect, I conceive there is a power or property applied to the wound, which hinders the irritation of contraction that naturally takes place in a wounded veffel, producing, probably, a paralyfis for a time.

tids. But in many fituations it will, probably, be impoffible to do this, with any hopes of fuccefs, and therefore we
may have recourfe to the fympathetic affections before defcribed.

Too much action, with fmall powers may often, if not
always, be claffed with the irritable conftitution, and bleeding fhould then be performed with very great caution : one
cafe out of many I fhall relate as an inftance of great action
with debility. A gentleman had one of the moft violent inflammations I ever faw, in one of his eyes attended with violent pain in his head, the blood extremely fizy, all of which
denotes great action of parts; yet the buff of the blood
was fo loofe when coagulated, that it could hardly bear
its own weight, or make any refiftance to the finger when
preffed; and although he was bled pretty freely, yet he never found any relief from it. This blood becoming a fymptom, both of the conftitution and difeafe, manifeftly fhowed
weak powers from its loofenefs, and too great action from
its flownefs of coagulation, which was the caufe of the
buff.

The following cafe is another ftrong inftance of great
action in a weak, irritable habit. A lady had a violent inflammation at the root of the tongue, fo as to form a confiderable fuppuration; with a pulfe of one hundred and
twenty, one hundred and twenty-five, and often one hundred and thirty, in a minute : her blood was extremely
fizy, yet fhe received but little benefit from the firft bleeding, although the blood coagulated pretty firmly, which
indicated ftrength. She was of an irritable conftitution,
fo as to receive lefs benefit from bleeding than another;
and when bled three times, the blood became extremely
loofe in its texture, which bark removed, as well as the
other fymptoms. Upon leaving off the bark, the fymptoms all recurred, and when fhe was bled again for the
fecond attack, which was the fourth time the blood, although inflammatory, had recovered a good deal of its proper
firmnefs; but in the fecond bleeding, for this fecond attack,
it was lefs fo; and in the third it was ftill lefs. Sufpect-
ing that bleeding in the prefent cafe would not produce refolution, I paid particular attention to the pulfe at the time
of bleeding, and found that in this laft bleeding the pulfe
increafed in its frequency even in the time of bleeding; and
within a few minutes after the bleeding was over, it had

increased ten strokes in the minute*. These bleedings retarded suppuration, but by producing irritability they could not effect resolution.

Where there is a disposition to form but little blood, when known, bleeding should be performed with great caution.

When the inflammation is far from the source of the circulation, the same precautions are necessary. In general it can be taken away from the part in such cases. But these are only so many facts, that require peculiar symptoms to ascertain them.

The common indications of bleeding, besides inflammation, are too often very little to be relied upon; and I shall consider them no further than as it concerns inflammation; which will indeed throw light on other cases. The pulse is the great indication in inflammation; but not always to be depended upon.

In inflammations that are visible, a knowledge of the kind of inflammation is in some degree ascertained, as has been observed, we therefore go upon surer ground in our indications for bleeding: but all inflammations are not visible; and it is, therefore, necessary to have some other criterion: however, if we could ascertain the pulse, peculiar to such and such appearances, in visible inflammation, and that was universally the same in all such appearances, we might then suppose that we had got a true indicative criterion for our guide, and therefore apply it to invisible inflammation, so as to judge of the inflammation by the state of the pulse; but when we consider, that the same kind of inflammation in every part of the body will not produce the same kind of pulse, but very different kinds, not according to the inflammation, but according to the nature of the parts inflamed, and those other parts also not visible, we lose at once the criterion of pulse as a guide. When we consider, also, that there shall be every other sign, or symptom of inflammation in some viscus, and from the symptoms the viscus shall be well ascertained, yet the pulse shall be soft,

---

* This, of the pulse increasing upon bleeding, is not always to be set down as a sure sign of irritation being an effect; for in a sluggish pulse, arising from too much blood, the increase of stroke, and freedom given to the circulation is salutary; but when a pulse is already quick, an increase must arise from irritation.

and of the common frequency; and upon bleeding, in consequence of thefe inflammatory fymptoms, the blood fhall
correfpond exactly with all of them, except the pulfe, it
fhall be fizy, firm, and cup, as was the cafe in a lady,
which has been before defcribed, we fhall be ftill farther
convinced that the pulfe is a very inadequate criterion,   If
a pulfe be hard, pretty full, and quick, bleeding appears
to be the immediate remedy, for hardnefs rather fhews
ftrong contractile action of the veffels not in a ftate of inflammation, which alfo implies ftrong action of the blood ;
and from fuch a pulfe, a fizy blood will generally be found;
but even a quick, hard pulfe, and fizy blood, are not always to be depended upon as fure indications of bleeding
being the proper method of refolution of inflammations ;
more muft be taken into the account.

The kind of blood is of great confequence to be known ;
for although it fhould prove fizy, yet if it lies fquat in the
bafon, and it is not firm in texture, and if the fymptoms,
at the fame time, are very violent, bleeding muft be performed very fparingly, if at all ; for I fufpect that under
fuch a ftate of blood, if the fymptoms continue, bleeding
is not the proper mode of treatment.   The cafes of this
kind, which have been related, are ftrong proofs of this.

As the pulfe abftracted from all other confiderations,
is not an abfolute criterion to go by, and as fizy blood, and
a ftrong coagulum are after proofs, let us fee if there are
any collateral circumftances that can throw fome light on
this fubject, fo as to allow us to judge, à priori, whether it
be right to bleed or not, where the pulfe does not of itfelf
indicate it.   Let us remember, that in treating of inflammation of different parts, I took notice of the pulfe peculiar to each part, which I may now be allowed to repeat.
Firft, I obferved that an inflammation in parts not vital,
or fuch as the ftomach did not fympathize with, if there
were great powers, and the conftitution not very irritable,
the pulfe was full, frequent, and hard.   Secondly, that
on the contrary, in inflammations of the fame parts, if the
conftitution was weak, irritable, etc. then the pulfe was
fmall, frequent and hard, although perhaps not fo much fo
as when in vital parts.   Thirdly, that when the inflammation is in a vital part, fuch as the ftomach, inteftines, or
fuch as the ftomach readily fympathizes with, then the
pulfe is quick, fmall, and hard, fimilar to the above. Now,

in the firft ftated pofitions we have fome guide, for in the firft of thefe, viz. where the pulfe is ftrong, etc. there bleeding is moft probably abfolutely neceffary, and the fymptoms, with the ftate of blood joined, will determine better the future conduct ; but in the fecond, where the pulfe is fmall, very frequent, and hard, bleeding fhould be performed with great caution ; yet in inflammation of the fecond ftated parts, the conftitution feems to be more irritable, giving more the figns of weaknefs, as if lefs in the power of the conftitution to manage.

Bleeding, reftricted to two or three ounces, can do no harm, by way of trial; and, as in the firft cafe, the fymptoms and blood are to determine the future repetition ; but in the third, or vital parts, viz. either the ftomach, or fuch as the ftomach fympathizes with, we are yet, I am afraid, left in the dark refpecting the pulfe. Perhaps, bleeding at firft with caution, and judging from the blood and its effects upon the other fymptoms is only the criterion we can go by.

The kind of conftitution will make a material difference, whether robuft, or delicate.

The mode of life will alfo make a material difference, whether accuftomed to confiderable exercife, and can bear it with eafe : conftitutions fo habituated will bear bleeding freely, but thofe with contrary habits will not. The fex will likewife make a difference, although the mode of life will increafe that difference; therefore men will bear bleeding better than women: even age makes a material d ff rence, the young being able to lofe more blood than the old ; for the veffels of the old are not able to adapt themfelves fo readily to the decreafed quantity; it even fhould not be taken away fo quickly ; and prob bly the conftitution may, in fome degree, have loft the habit of making blood, fince it has loft the neceffity.

The urine will throw fome light on the difeafe ; if high coloured, and not much in quantity, it may be prefumed, with the other fymptoms, that bleeding will be of fingular fervice ; but if pale, and a good deal of it, although the other indications are in favour of bleeding, yet it may be neceffary to do it with caution.

However, bleeding fhould in all cafes be performed with great caution, more particularly at firft ; and no more taken than appears to be really neceffary; it fhould only be done to eafe the conftitution, or the part, and rather lower it where the conftitution can bear it : but if the conftitution is already below or brought below a certain point, or gives the figns of it from the fituation of the difeafe, then an irritable habit takes place, which is an increafed difpofition to act without the power to act with. This, of itfelf, becomes a caufe of the continuance of the original difpofition, and therefore will admit neither of refolution, nor fuppuration, but continue in a ftate of inflammation ; which is a much worfe difeafe, than the former.

Upon any other principle than thofe abovementioned, I cannot fee why bleeding fhould have fuch effects in inflammation as it fometimes has. If confidered in a mechanical light, as fimply leffening the quantity of blood, it cannot account for it ; becaufe the removal of any natural mechanical power, can never remove a caufe which neither took its rife from, nor is fupported by it : however, in this light it may be of fome fervice; becaufe all the actions relative to the blood's motion will be performed with more eafe to the folids, when the quantity is well proportioned.

It is probably from that connexion between the folids and fluids, that the conftitution, or a part, is in a ftate of perfect quietude, or health, in which we find that the fluids are, and ought to be, in a large quantity ; but in a ftate of inflammation, or increafed powers and actions, thofe proportions do not correfpond, at leaft in the parts inflamed ; and by producing the equilibrium between the two, fuitable to fuch a ftate, the body becomes fo far as this one circumftance can affect it, in a ftate of health ; and this in many cafes will caft the balance in favour of health : it is not, however, fufficient to produce this effect in all inflammations.

How far taking the blood from parts peculiarly fituated with refpect to the parts inflamed, is more efficacious, I believe is not yet determined ; as bleeding in the left fide for an inflammation in the right, upon the fuppofed principle of derivation, which might be claffed with remote fympathy ; but fo far as the lofs of the blood acts mechanically, viz. fo far as it fimply empties the veffels, it certainly can have no more effect than if taken in any other way ; nor can it affect the living principle, either univerfally, or locally, more in this mode than in any other ; but how far it can affect the fympathizing principle, I do not know.

Bleeding is often performed from no conftitutional indication, but only as a preventive, arifing from experience ;
fuch as in confequence of confiderable accidents, as blows
on the head, fractured bones, etc. but this is not to the prefent purpofe.

XI. THE USE OF MEDICINES INTERNALLY, AND OF LOCAL APPLICATIONS IN INFLAMMATION.

Every thing given to the body, or applied to the part
inflamed, that can abate inflammation, or its effects in the
conftitution, may be called medicine ; fuch therefore, divide themfelves into conftitutional, and local : the firft
will be internal, the fecond external ; but whichfoever
way they are applied, they that tend to leffen inflammation,
have their effects local ; for mercury, although given internally, for a venereal ulcer in the throat, yet acts locally
on the difeafe ; but thofe that tend to remove conftitutional
affections, have their effects conftitutional.

The internal medicines generally ordered for the refolution of inflammation, are fuch as tend to have fimilar effects to that which is produced by bleeding ; namely, lowering the conftitution, or the action of the parts ; and this
has ufually been performed principally by purges ; and the
medicines that were given to remove, or leffen the effects
of inflammation on the conftitution, have been fuch as generally tend to leffen fever, or the effects that the inflammation has upon the conftitution.

Purges were generally given in cafes of inflammation,
probably at firft from the idea of humors to be difcharged)
and fuch practice will anfwer beft where bleeding fucceeds ; becaufe it will lower the body to a more natural
ftandard, and of courfe the inflamed part, as a part of that
conftitution ; but here the fame cautions are neceffary that
were given upon bleeding, becaufe nothing debilitates fo
much as purging, when carried beyond a certain point.
One purging ftool fhall even kill, where the conftitution is
very much reduced, as in many dropfies : therefore, keep-

ing the body fimply open, is all that fhould be done.
However, although purging lowers confiderably, yet its effect is not fo permanent as that of bleeding ; it rather lowers action, without diminifhing ftrength for if a perfon
was to feel the lofs of blood equal to a purge, that fenfation would be more lafting.

Many conftitutions rather acquire ftrength upon being
gently purged ; particularly fuch as have been living above
par ; but fuch ftrength as is acquired by putting the body
in good order, I fhould fuppofe is not inimical to inflammation.

In irritable habits, where the inflammation becomes
more diffufed, greater caution is neceffary, with regard to
purging, as well as bleeding ; for I obferved on the fubject of bleeding, that in fuch conftitutions, no more blood
fhould be taken than would relieve the conftitution, as it
were mechanically, but not fuch a quantity as to have a
tendency towards lowering or weakening that conftitution ;
for in fuch cafes the action is greater than the ftrength; and
whenever the difpofition between thefe two is of this kind,
we cannot expect any thing falutary from this mode of
treatment, and therefore fhould not increafe it.   In fuch
cafes, the very reverfe of the former method fhould often be
practifed ; whatever has a tendency to raife the conftitution
above irritability, fhould be given ; fuch as bark, etc. The
object of this laft practice confifts in bringing the ftrength
of the conftitution, and part, as near upon a par with the
action as poffible, by which means a kindly refolution, or
fuppuration, may take place, according as the parts inflamed are capable of acting.

Medicines, which have the power of producing ficknefs,
leffen the action, and even the general powers of life, for
a time, in confequence of every part of the body fympathizing with the ftomach, and their effects are pretty quick.

Ficknefs lowers the pulfe ; makes the fmaller veffels contract, and rather difpofes the fkin for perfpiration, but
not of the active or warm kind ; but I believe it fhould
proceed no farther than ficknefs ; for the act of vomiting
is rather a counteraction to that effect, and produces its
effects from another caufe, and of courfe of another kind,
which I believe rather roufe : it is probably an action arifing from the feel of weaknefs, and intended to relieve the
perfon from that weaknefs. It is fimilar to the hot fit of

an ague; a counteraction to the cold one. There are few
so weak, out they will bear vomiting, but cannot bear sick-
nefs long.

If we had medicines, which, when given internally,
could be taken into the conftitution, and were endowed
with a power of making the veffels contract, fuch I apprehend, would be proper medicines. Bark has certainly this
property, and is of fingular fervice, I believe, in every inflammation attended with weaknefs, and therefore, I conceive, fhould be oftner given than is commonly done; but
it is fuppofed to give ftrength which would not accord with
inflammations attended with too much ftrength and confiderable irritation,

Preparations of lead, given in very fmall dofes, might
be given with fuccefs, in cafes attended with great
ftrength.

Applications to the body to cure or refolve inflammations
are, with regard to their mode of application, of two
kinds; one is applied to the part inflamed, the other to
fome diftant part: the firft may be called local or abfolute,
refpecting the part itfelf; the fecond, relative; but even
the firft may be confidered as having a relative effect in one
of its modes of action, viz. that called repulfion, from
which local applications have by fome been objected to, and
it is principally local applications that can repel, although
not literally.

The firft, or abfolute effects of medicine, may be divided
into two kinds, viz. one, the fimple cure of the part; the
fecond, producing an irritation of another kind in the part;
both, however, act locally, and their ultimate effect is local. Local applications to a part, where that application
poffeffes really the powers of refolution, muft be much
more efficacious than any of the other modes of refolution;
for inftance, mercury has much greater powers when applied immediately to the venereal complaint, than when applied
to the neareft furface; where, however, we have not medicines that can refolve inflammation by application, then
of courfe the other method is the moft efficacious, but whether we have external or local applications which have really a tendency to leffen the inflammatory difpofition, is
not well afcertained. I doubt our being in poffeffion of

Vol. Ii. K

many that can remove the immediate cause.   Such would of course remove the action, or if not wholly, would at least lessen it, and allow the inflammation to go off.

But most of our powers in this way appear to be of the soothing kind, which, therefore, lessen the action, although the cause may still exist, and hence the effects are also lessened.   This either produces a termination of the inflammation, or it is protracted, the cause lessens, and the inflammation wears itself out.

As inflammation has too much action, which action gives the idea of strength, such applications as weaken have been recommended, and cold is one of them.   Cold, according to its degrees, produces two very different effects, one is the exciting of action without lessening the powers, the other is absolutely debilitating, while at the same time it excites action, if carried too far ; in the first, it becomes like suitable exercise to the vascular system, as bodily exercise is to the muscles, increasing strength ; but when carried or continued beyond this point, it lessens the powers, and becomes a weakner, calling up the action of resistance after the powers are lessened ; therefore cold should not be indiscriminately used, and should be well proportioned to the powers.

Cold produces the action of contraction in the vessels, which is an action of weakness.   A degree of cold suddenly applied, which hardly produces more than the sense of cold, excites action after the immediate effect is over, which is the action of dilatation, and which is the effect of the cold-bath when it agrees ; and as cold produces weakness in proportion to its degree, its application should not be carried too far, for then it produces a much worse disease, irritability ; or over action to the strength of the parts, and then indolence too often commences.   Cold might be supposed to act on an inflamed part, similar to its action on a frozen part, restraining action, keeping it within the strength of the part in the one case, so as not to allow death to take place from over action ; and in the other, to keep it within bounds*.

* As cold can be applied upon two very different principles, it is necessary to mention which is here meant.   When cold is applied either within the powers of resistance of the part, to excite heat; or only for so short a time as to give the stimulus of

Lead is alfo fuppofed to have confiderable effects in this way ; but I believe much more is afcribed to it than it deferves.

The property of lead appears to be that of leffening the powers and not the action, therefore fhould never be ufed but when the powers are too ftrong, and acting with too much violence : however, lead certainly has the power of producing the contraction of the veffels, and therefore where there is great ftrength, lead is certainly a powerful application.

Applications which can weaken fhould never be applied to an irritable inflammation, efpecially if the irritability arifes from weaknefs ; I am certain I have feen lead increafe fuch inflammations, particularly in many inflammations of the eyes and eyelids ; and I believe it is a bad application in all fcrofulous cafes ; in fuch cafes the parts fhould be ftrengthened without producing action.

Warmth, more efpecially when joined with moifture, called fomentation, is commonly had recourfe to ; but I am certain that warmth when as much as the fenfative principle can bear, excites action ; but whether it is the action of inflammation, or the action of the contraction of the veffels, I cannot determine ; we fee that in many cafes they cannot bear it, and therefore might be fuppofed to increafe the action of dilatation, and do hurt ; but if that pain arifes from the contraction of the inflamed veffels, then it is doing good, but this I doubt, becaufe I rather conceive the action of contraction would give eafe.

Acids have certainly a fedative power, as alfo alcohel, and I believe many of the neutral falts.

I believe it is not known that we have the power of adding ftrength to a part by local application ; that, in general, I believe muft arife from the conftitution; for although

cold, then a re-action takes place, and warmth is the confequence ; but if cold is applied beyond the powers of refiftance, then a contraction of the veffels takes place, and that contraction is in fome degree permanent ; but this muft be done with caution, for if continued too long, it will produce debility, and action will be excited which will be irritable. In the prefent, the application of cold fhould only be fufficient to excite the contraction of the veffels, and that not contained too long, for reafons above affigned.

we have the power of giving action, yet this does not imply ftrength.

Many local applications are recommended to us, refpecting many of which I have my doubts.

The mode of cure by an irritation different from the difeafe, appears to increafe the difeafe, but by deftroying the firft mode of action it produces another difeafe, viz. one according to the mode of irritation of the application, and which more eafily admits of a cure than the firft. I believe, however, that this takes place principally in fpecific difeafes, and not fo readily in common inflammation; for a common inflammation moft probably would be increafed by it. I have known fpecific inflammations much more eafily cured by their fpecific medicine, than the common inflammation of the fame conftitution, viz. I have feen a gonorrhœa and a chancre cured much more eafily in fome conftitutions, than an inflammation from an accident, and oftener than once or twice in the fame perfon. However, this mode will not do in all fpecific difeafes, for the fcorfula will not change its nature by it, nor will the irritable, although fpecific. The venereal gonorrhœa (if parts are very irritable) is an inftance of this, for irritating injections increafe it; ftill we have many cutaneous inflammations cured in this way; for a pretty ftrong folution of corrofive fublimate will remove an inflammation of the fkin. The unguentium citrinum mixed with any common ointment, cures many inflammations of the eyelids; yet I believe that artificial irritations are fimilar to one another; and I do not know if there be any difference between them, although I can conceive one may agree better with fome conftitutions than others. However, thefe local or immediate applications can only be fuch as come in contact with the difeafe, which always muft be fome expofed furface, as when the fkin of the eyelids, tonfils, etc. are inflamed; but even there fome part muft be affected by continued fympathy, if they produce a cure, as the inflammation generally goes beyond the furface of immediate contact.

That inflammation which admits of repulfion, although by local means, might be confidered here but from its effects and connexion with the conftitution, it comes in more properly with the feveral relations, under which I fhall confider it,

XII. GENERAL OBSERVATIONS ON REPULSION, SYMPATHY, DERIVATION, REVULSION, AND TRANSLATION.

THESE terms are meant to be expreffive of a change in the fituation of difeafed actions in the body, and they are fo named according to the immediate caufe ; for any one difeafe may admit of any of thefe modes equally ; that is, a difeafe which admits of being repelled, may admit of being cured by fympathy, which probably includes derivation, repulfion, and tranflation. That fuch a principle or principles exift, is, I think evident ; but the precife mode of action is, I believe, not known ; that is, it is not known what part of the body more readily accepts of the action of another ; if there are fuch parts, they might be called correfpondent parts, whether the action changes its place from repulfion, fympathy, derivation, or tranflation. In derivation and repulfion, whether one mode of irritation is better than another, to invite or divert the action, and whether parts of a peculiar action do not require a peculiar irritation to divert them ; to all this we are likewife totally ftrangers.

It is not to my prefent purpofe to go into the different effects of this principle; although I muft own it might be as ufeful a part of the healing art as any ; and even more, for it is probably the leaft known, as being the leaft intelligible, and therefore the more ufe may be derived from its inveftigation.

The operations denominated by thefe terms, fo far as they exift, appear all to belong to the fame principle in the animal œconomy, for they all confift either in a change of the fituation of a difeafe or its action ; a change of the fituation, as in the gout, a change of the action, as a fwelling of the tefticles in the ftopping of a gonorrhœa. This laft is not properly a change of the fituation of the difeafe, but only of the general inflammatory action without the fpecific action, thefe principles can only produce a change in the feat of action; not in any of the confequences of difeafe ; they have in fome inftances a connexion with the natural operations of the body, as it were interfering with them ; and when

that is the cafe, they in general muft produce difeafe of
fome kind. Thus, the ftopping of the menfes, a local natural action, arifing from the conftitution, which may be
effected by local applications, called repelling, by a derangement of the conftitution, and by many confequences
which depend on a deranged conftitution fimply, or it may
be drawn off by a derangement of the conftitution, which is
a kind of derivation or revultion. We find that local applications derange alfo other parts, which have no vifible effect upon the part of application as the above, nor any vifible connexion with the parts which affume the action.
Thus cold, efpecially if wet be applied to the feet, will
bring on complaints in the ftomach and bowels, by fympathy; and the fame mode of application of cold, if local, will
produce a local complaint : as cold air blowing on a part
will bring on rheumatifm.

Thefe changes were all fuppofed formerly to be of more
confequence than I apprehend they really are; for they are
only the change of fituation of difeafe. They were introduced into the œconomy of difeafe from the idea of humors.
Repellants were fuch applications as drove the humours out
of a part, which would fall on fome other ; fympathy con-
fifted in another part taking them up ; derivation was a
drawing off, or inviting the humors; revulfion was the fame,
and tranflation was the moving of humors from one part to
another. Thus we have thofe different terms applied to
that connexion of parts, by which one part being affected,
fome other is affected or relieved ; or, as in tranflation,
fome other part takes up the difeafe as it were voluntarily,
as is often the cafe in the gout. All of thefe produce one
of the fymptoms of a difeafe, viz. fenfation and inflammation ; but I believe feldom or ever real difeafed ftructures.
This agrees with what was formerly obferved, that local
inflammations, depending on the conftitution, feldom or e-
ver fuppurate.

I believe that thefe powers have greater effects in difeafes, depending on, or producing action and fenfation, which
are called nervous, than on thofe producing an alteration
in the ftructure of parts.

Thus, we have the cramp in the leg cured by a gentle
irritation round the lower part of the thigh, fuch as a gar-
ter, which may be faid to arife from derivation, or fympathy,

I have known in a nervous girl, a pain in one arm cured by rubbing the other.

Thefe cures by derivation, repulfion, tranflation, etc. do not deferve that name, although the patients are cured of the original difeafe, as in many cafes there is as large a quantity fomewhere elfe in the body uncured : for example, in thofe cafes where the cure is from fome local inflammation being produced, and perhaps more violent than the firft ; but in other cafes where the cure arifes only from an action in a part without a difeafed alteration of ftructure, then the cure, in fuch cafes, is performed without any other difeafe having been produced ; fuch as ficknefs or vomiting, curing a difeafe of the tefticles.

I have already obferved that local applications were principally fuppofed to repel, by the firft or fecond mode of action ; yet internal medicines having a fpecific, or what might be called a local action, although given internally, may repel by ftopping the difeafed action in the part which it chiefly affects, fuch as mercury falling on the mouth, might repel a difeafe from the mouth. Hemlock might do the fame, with regard to the head ; or turpentine with regard to the urethra. In the laft, we often find by taking balfams, that by ftopping the difcharge, a fwelling of the tefticles comes on, or an irritable bladder. As repulfion in this way is not fo evident, it is lefs noticed. The uncertainty in the power of medicines, refpecting repulfion, has led furgeons into more errors than any other principle in the animal ceconomy, with regard to difeafes. It has prevented their acting in many cafes, where they might have done it with fafety and effect. A ftronger inftance cannot be given than in that fpecies of the venereal difeafe, called a gonorrhœa, which they did not venture to cure by local applications, for fear of driving it into the conftitution, and producing a pox ; but they did not confider that a gonorrhœa did not arife from the conftitution, but may be faid to arife from accident, or at leaft is entirely local, and therefore no repulfion could take place. The idea of repelling was firft introduced when local difeafes were fuppofed to arife from a depofit or derivation of humors to a part, and is ftill retained by thofe who cannot or will not allow themfelves to think better ; yet ftill the term might be applied to difeafed action, for the removal of many difeafed actions from a part which fall on fome other part, is certainly the repelling of that difeafed action ; but fince it is not fubdued, but only driven from the part, as is often the cafe with the gout, no cure is performed by this means.

Both or either of the two local methods of removing difeafe, juft now mentioned, viz. whether by fimply curing the difeafe, or by deftroying the difeafed action, in confequence of exciting an action of another kind, may produce the effect called repulfion ; but the former, I believe, can only take place in inflammations arifing from the conftitution, and which being prevented from fettling in this part, return upon the conftitution again, and often fall upon fome other part, viz. one next in order of fufceptibility for fuch inflammation, as is often the cafe in the gout, and in many other difeafes befides inflammation, as in many nervous complaints. St. Vitus' dance is a remarkable inftance of it; but in this cafe it is not to be confidered as a cure of the difeafe, but only as a fufpenfion of its action in this part.

I could conceive it poffible that the fecond mode of local cure, which is by producing an irritation of another kind, might not repel ; although it cured the firft or local complaint, becaufe there is in fuch modes of cure ftill a larger quantity of inflammation in the part than was produced by the difeafe (although of another kind) ; but as the idea of repelling is to have a difeafe fomewhere, although not in the fame place, keeping it in the prefent fituation may be as proper, if not more fo, than in any other it might go to. But if, on the other hand, the conftitution requires to have a local complaint arifing from itfelf, which, as it were, is drawing off, or relieving this conftitutional difpofition, then curing the one already formed, by producing another in the fame part, can be of no fervice ; for if the artificial difeafe is not of the fame nature with the conftitutional one, (which it cannot be) if it deftroys the other, then it cannot act as a fubftitute for the other. We may obferve that by producing an irritation of another kind in the gout, we may deftroy the gouty inflammation in the part, but cannot always clear the conftitution of it ; there is, therefore, no benefit arifing from fuch practice in thefe cafes.

The repelling powers which act from applications being made to the parts immediately affected, or by the change of one difeafe into another, become the moft difficult of any to be afcertained ; becaufe it muft be very difficult to fay, what will merely repel and what will completely cure, or perfectly change the difeafe. Repulfion is certainly to be confidered as a cure of the part, whatever may be the confequence; and a change in the difeafe is certainly a cure of the firft, although a difeafe may ftill exift in the part.

That an artificial irritation made on one part does not (always at leaft) cure or remove a difeafed irritation of a fpecific nature in another part, is, I think, evident in many cafes, even although that fpecific fhould be an affection of the conftitution. This, at leaft, was evident in a cafe of the gout, for when the gout was in fome of the vital parts, and finapifms were applied to the feet, they did not relieve thofe vital parts: although the inflammation in confequence of the finapifms was confiderable; but this inflammation brought on the gout on the feet; and as foon as this happened, the vital parts were relieved; from which it would appear, that a fpecific irritation required a fpecific derivator. It may be fuppofed that the inflammation, in confequence of the finapifms, brought on, or produced, fuch a derangement in the feet as made them more fufceptible of the gout; or the inflammation became an immediate caufe of the action of gout taking place there.

It is plain too, that where there is a gouty difpofition, or a gouty action, in the coftitution, a derangement in a part may bring it on; for in the above perfon, who had ftill thofe internal fpafms recurring upon the leaft exercife or anxiety of mind, but was in all other refpects, and at all other times well, by applying finapifms a fecond time to his feet, till a confiderable cutaneous inflammation came on, the gout attacked the ball of the great toe of the right foot, and the laft joint of the great toe of the left, which lafted about two days. This attack of gout, however, did not relieve him of the remaining fpafms, as the firft did; and therefore was to be confidered an additional gouty action. This could certainly not have taken place if the conftitution had not been gouty.

In difeafes where we have no fpecific application capable of acting immediately, the advantages gained by derivation, revulfion, or fympathy, are much greater in many cafes than by the effects of any local application at prefent

known ; and the medicines which are capable of producing this effect are often such as would either have an effect if applied to the diseased part, or would increase it. This
arises from the dissimilar actions of the two parts ; that is,
the diseased actions of the one being similar to, or producing the actions of recovery in the other ; nor is it difficult
to conceive why it should be so ; for since the medicines are
not specifics, but only invite or remove the disease by that
connexion which the living powers of the one part have
with those of another, it is reasonable to suppose that this
principle of action between the parts must be much stronger
than the effects of many medicines which have only a tendency to cure ; or perhaps no tendency that way at all.
Thus we find, that vomits will often cure inflammations of
the testicles, when all soothing applications prove ineffectual, and when the same emetic could not have the least
effect on the part itself, were it applied to it.

Thus, we also find that a caustic behind the ear will relieve inflammations of the eyes or eyelids, when every application to the parts affected has proved ineffectual, and when
this caustic, if applied to the parts themselves only as a sti-
mulent, would increase the disease.

Sympathy, perhaps, (except the continued) includes the
mode of action in all of those which I have called relative,
viz. repulsion, derivation, revulsion, and translation ; at least
it is probably the same principle in the whole. What I would
call a cure by sympathy, is producing a curative action in
a sound part, that the diseased may take on the same mode
of action from sympathy, that it would take on, if the curative was applied to it ; so that sympathy might even be
supposed to repel in cases which would admit of repelling,
and fall on some other part, although not the part necessa-
rily where the application was made. The difference between derivation, or repulsion, and sympathy, consists in
derivation producing a disease in a sound part to cure a disease in another part, as was observed ; while sympathy
is applying the cure to a sound part to cure the diseased ;
but in many cases it will be very difficult to distinguish the
one from the other.

Sympathy is very universal, or more general than any
others ; for there are few local diseases that do not extend
beyond the surface of contact, which produces continued

fympathy'; and also, there are few parts that have not fome connexion with fome other part, which gives us remote fympathy.

It may be recollected, that when fympathy was treated of, it was divided into continued, contiguous, remote, fimilar, and diffimilar.

The cure by continued fympathy is that application which we have reafon to fuppofe would cure if applied to the part itfelf; fuch as applying mercury to the fkin over a venereal node. The node is cured by its fympathizing with the mercurial irritation of the fkin; and the action of the fympathizer here is fimilar to the action of the part of application. Remote fympathy is feldom or ever produced by a fimilarity of action in fimilar parts; but moft probably cures by diffimilar modes of action in the two parts, and therefore might be called diffimilar fympathy, viz. by ftimulating the part of application in fuch a way that the fympathizer fhall act in the fame way as if the real application of cure was made to it, and yet the mode of action of the part of application fhall not be at all fimilar to the fympathizer. I can even fuppofe a local difeafe cured by fympathy and by that medicine which would increafe it, if applied immediatly to it. Let us fuppofe, for example, any difeafed mode of action, and that this mode could be increafed by fome irritating medicine, if ppiied to it; but apply this irritator to fome other part which this difeafed part fympathizes with, and that the fympathetic act in the difeafed part fhall be the fame as if its curative medicine was applied to it, fimilar to what would have taken place, if its fpecific irritator was applied; then, in fuch a cafe the medicine would cure by fympathy, although it would increafe the difeafe if applied locally, or have no effect at all.

The contiguous fympathy is where it would appear to act from the approximation of diffimilar parts, and therefore is not continued fympathy; neither can it be called remote fympathy, as there appears to be no fpecific connexion, but to arife entirely from contiguity or approximation of parts. Of this kind are blifters on the head, curing headach; on the cheft, curing pains in the cheft; to the pit of the ftomach, to cure irritations there; to the belly, to cure complaints of the bowels.

The applications which act by contiguous fympathy are only thofe which can be applied to the neareft furface to that which is inflamed, and the inflamed part beyond this furface becomes affected in fome degree, fimilar to the part of application, fuch as the applications to the eyelids, when it is in the eye ; to the fcrotum, when in the tefticle ; to the abdomen, when fome of the bowels are inflamed ; to the thorax, in inflammaticns of the lungs etc.

Thefe may be either of the fpecific, ftimulating, or foothing kinds, fomething which affects the parts in fuch a manner as that a remote difeafed action ceafes. It may be fpecific, as opium applied to the pit of the ftomach curing an irritation of that vifcus ; ftimulating, as blifters to cure inflammation in the fubjacent vifcera, as has been mentioned ; foothing, as fomentations to the abdomen to relieve complaints in the bowels.

Derivation means a ceffation of action in one part, in confequence of an action having taken place in another ; and when this is a ceffation of a difeafed action, then a cure of that action in the original part may be faid to be performed ; this cure was brought into ufe from the idea of humors ; that is, the drawing off of the humors from the feat where they had taken poffeffion ; but I believe much more has been afcribed to it than it deferves.

How far it really takes place, I have not been able fully to afcertain in all its parts ; that is, how far the real difeafe is invited, and accepts of the invitation ; but I have already obferved, that there is fuch a principle of difeafe in the animal œconomy, although we muft fee from derivation, that the fame quantity, or perhaps more irritation is retained in the conftitution ; yet the artificial irritation produced being either fuch as more readily admits of a cure than the difeafed part, or being in parts which are not fo effential to life, an advantage by this means is gained ; thus burning the ear is practifed as a cure for the tooth-ach, and the part which is burnt more readily admits of a cure, than the tooth.   We alfo find that blifters often cure or remove deep feated pains, fuch as head-ach, and relieve the bladder when applied to the perineum.   Blifters and cauftics behind the ear, cure alfo inflammations of the eye.

Lefs may be faid on revulfion, fince we have explained derivation.

To draw off a difeafe always implies fafe ground, and can be applied with fafety in any difeafe : revulfion can beft be applied, when the difeafe attacks effential parts where the application cannot be fo near as to imply derivation.

Thus we find that vomits will cure an inflammation of the tefticles, white fwellings, and even venereal bubos ; and finapifms applied to the feet relieve the head.

Tranflation differs from derivation, revulfion, and repulfion, only as it proceeds from a natural or fpontaneous caufe, whereas thefe proceed from an artificial, accidental, or external caufe, and the common principle of them all feems to be fympathy ; for if not an act of its own, then it muft be either repelled, derived, or from fympathy.

Very ftrange inftances of tranflation are given us ; it has been fuppofed, that pus already formed has been tranflated to another part of the body depofited there in form of an abfcefs, and then difcharged ; this is an operation abfolutely impoffible ; matter abforbed may be carried off by fome of the fecretions, fuch as by the kidnies, which have a power of removing more than they fecrete ; but the depofition of pus is the fame with its formation.

Both revulfion and repulfion may be reckoned a fpecies of tranflation.

The gout moving of its own accord from the ftomach to the foot, or from one foot to the other, may be reckoned a tranflation of the gout.

XIII.   OF THE DIFFERENT FORMS IN WHICH MEDICINES ARE APPLIED, AND THE SUB-SIDING OF INFLAMMATION.

FOMENTATIONS, or fteams, wafhes, and poultices, are the common applications to a part in the ftate of inflammation. The firft, and laft, are commonly applied to inflam-

mation arising from external violence, and proceeding to
suppuration ; the second, commonly to internal surfaces,
as the mouth, nose, urethra, vagina, rectum, etc.   The
action of the two first is but of very short duration.

Fomentations and steams, are fluid bodies raised into vapor : they may be either simple, or compound ; simple,
as steam, or vapor of water ; compound, as steam of water impregnated with medicines.

This mode of applying heat, and moisture, appears from
experience to be more efficacious than when these are applied in the form of a fluid ; it often gives ease at the time
of the application, while at other times it gives great pain ;
but if it does give ease, the symptoms generally return
between the times of applying it, and with nearly the same
violence.   How far the application of a medicine for fifteen
minutes out of twenty-four hours can do good, I am not
certain : we find, however, that the application of a vapor
of a specific medicine, though but for a few minutes in the
day, will have very considerable effects : fumigations with cin-
naber, may serve as an instance.   The fomentations are
commonly composed of the decoction of herbs ; sometimes
the marsh mallows, etc. but oftner of the decoction of
herbs possessing essential oil, which I believe are the best,
because I suppose that whatever will excite contraction of
the vessels, will in some degree counteract the dilating
principle : vinegar, as well as spirits, are put into it ; how
far they stimulate to contraction, I do not know, but rather believe they remove irritation, which must lessen the
inflammatory action.

Washes are in general fluid applications, and are more
commonly applied to inflammations of internal surfaces,
than of the common integuments : there are washes to the
eye, called collyria ; washes to the mouth and throat, called gargles ; washes to the urethra, called injections ; and
to the rectum, called clysters ; but I am fearful that we
are not yet acquainted with the true specific virtues of
these washes, at least there is something very vague in their
application.   There are, for instance, astringent washes
for the inflammation of the eye, such as white and blue
vitriol, alum, etc. stimulating warm gargles for inflammations of the throat, such as mustard, red port, claret with
vinegar and honey ; but to moderate or resolve external
inflammations, they do not apply substances with any such

properties. How abfurd would it appear to furgeons in
general, if any one made ufe of the fame application to an
inflammation in any other part ; yet I do not know if there
is any difference between an inflammation of the eye or
throat, and one in any other part, if the inflammations are
of the fame kind : mercury cures a venereal inflammation
either of the eye or throat, as eafily as a venereal inflammation any where elfe, becaufe it is an inflammation of the
fame kind.

Thefe applications, like fomentations, are of fhort duration, for there is no poffibility of applying thefe powers
conftantly, except in the form of a poultice, whofe operation is fomewhat fimilar ; and, indeed, they are only fub-
ftitutes for a poultice, where that mode of application cannot be made ufe of, as I obferved with refpect to internal
furfaces.

Poultices are conftant applications, and like fomentations
may be two kinds, either fimply warm and wet, or medicated. The greateft effect that a poultice can produce
muft be immediate, but its power will extend beyond the
furface of contact, although only in a fecondary degree.

To the common inflammation, the fimpleft poultice is
fuppofed to be the beft, and that effect I believe is only by
keeping the parts eafier under the complaint ; but I am of
opinion, that fuch do not affect the inflammation any other
way. A common poultice is, perhaps, one of the beft applications when we mean to do nothing but to allow nature
to perform the cure with as much eafe to herfelf as poffible.

Poultices may be medicated, fo as to be adapted to the
kind of inflammation ; fuch as the folution of lead, opium, mercury, etc. in fhort, they may be compounded with
any kind of medicine.

Whatever the difpofition is, which produces inflammation, and whatever the actions are which produce the effects, that difpofition under certain circumftances, viz.
when it arifes either from the conftitution or the parts, can
be removed, and of courfe the actions excited by it. The
difpofition for inflammation fhall have taken place, and the
veffels which are the active parts, fhall have dilated, and allowed more blood to enter them, fo that the part fhall look
red, but no hardnefs or fullnefs fhall be obferved, and the

whole shall subside before adhesions have been formed ; or
if inflammation has gone so far as to produce swelling,
which is the adhesive stage of the disease, it by certain methods can be frequently so assuaged as to prevent suppuration taking place, and then the parts will fall back into
their natural state, which is called resolution ; some adhesions being perhaps the only remaining consequences of
the inflammation.

The same methods are likewise often used with considerable success in lessening inflammation arising from violence,
so as to prevent suppuration entirely ; but in many of these
cases they are not sufficient, and in those where it cannot
be prevented, yet it may be lessened by the same means.

As the first symptom of inflammation is commonly pain,
so is the first symptom of resolution a cessation of that pain,
as well as one of the symptoms of suppuration, which is a
species of resolution. I have known the cessation of pain
so quick, as to appear like a charm, although no other visible alteration had taken place, the swelling and colour being the same.

Why inflammation of any kind should cease after it has
once begun, is very difficult to explain, or even to form an
idea of, since yet we have no mode of counteracting the
first cause or irritation ; it may be supposed to arise from
the principle of parts adapting themselves in time to their
present situation, which I call custom, and that therefore in
order to keep up the inflammation, it would be necessary
for the cause to increase, in proportion as the parts get re-
conciled to their present circumstances ; but allowing this
to be the cause, it will not account for their returning back
to their natural or original state, when this increase of irritation ceases, and only the last or original irritation remains;
for upon this principle, they only grow more easy under
their present state ; or perhaps, which is worse, acquire a
habit of it, which may be the cause of many indolent specific diseases.

If we suppose the removal of the original cause to be
sufficient to stop the progress of inflammation, and when
this is stopped, that the parts cannot easily remain in the
same inflamed state, but by their own efforts begin to restore themselves to health ; which we can easily conceive
to be the case in the specific diseases, especially those arising

from poifons of fuch kinds as to be capable of a termination as the fmall pox ; or where a cure can be adminiftered for the effects of the poifon, as in the lues venera ; then we muft conclude that the inflamed ftate is an uneafy ftate, a force upon the organs which fuffer it ; like the bending of a fpring, which is always endeavouring to reftore itfelf, and the moment that the power is removed, returns back to its natural ftate again ; or it may be like the mind, forgetful of injuries.

XIV. OF THE USE OF THE ADHESIVE IN-FLAMMATION.

THIS inflammation may be faid in all cafes to arife from a ftate of parts in which they cannot remain, and therefore an irritation of imperfection takes place. It may be looked upon as the effects of wife counfels, the conftitution being fo formed as to take fpontaneoufly all the precautions neceffary for its defence; for in moft cafes we fhall evidently fee that it anfwers wife purpofes.

Its utility may be faid to be both local and conftitutional, but certainly moft fo with regard to the firft. Its utility is moft evident when it arifes from a difeafe in a part, whether this proceeds from the conftitution or otherwife, and when it does, it muft be confidered as arifing from a ftate in which that part cannot exift, as in expofure, and therefore is the firft ftep towards a cure. It is often of fervice in thofe cafes which arife from violence, although not fo neceffarily fo, the injured parts not being always under the neceffity of of having recourfe to it, as was fhewn in treating of union by the firft intention.

When the adhefive inflammation arifes from the conftitution, it may depend on fome difeafe of that conftitution ; and if fo, it may be conceived to be of ufe to it,

especially if it should be supposed to be the termination of
an univerfal irritation in a local one, by that means relieving the conftitution of the former, as in the gout ; but
when it is only the fimple adhefive inflammation that takes
place, I am rather apt to think that it is more a part of
the difeafe, than a termination of it, or an act of the conftitution.

The adhefive inflammation ferves as a check to the fuppurative, by making parts, which otherwife muft infallibly
fall into that ftate, previoufly unite, in order to prevent its
accefs, as was defcribed in the adhefive inflammation being limited ; and where it cannot produce this effect, fo
as altogether to hinder the fuppurative inflammation itfelf
from taking place, it becomes in moft cafes a check upon
the extent of it.   This we fee evidently to be the cafe in
large cavities, as in the tunica vaginalis after the operation
of the hydrocele ; for after the water has made its efcape,
parts of the collapfed fack frequently unite to other parts
of the fame fack by this inflammation, and thereby preclude
the fuppurative from extending beyond thefe adhefions,
which fo far prevents the intention of the furgeon from
having its full effect : and often on the other hand, the
adhefive ftate of the inflammation takes place univerfally in
this bag, in confequence of the palliative cure, which produces the radical, and thereby prevents a relapfe. In the
hernea it performs a cure by uniting the two fides of the
fack together, by means of flight preffure, fo that we
fhould underftand perfectly its mode of action, where it
can prevent a cure, and where it can perform one. In
ftill larger cavities, fuch as the abdomen, where often only
a partial inflammation takes place, as is frequently the cafe
after child-bearing, and in wounds of this cavity, we find
this inflammation produced, which either prevents the fuppurative altogether, or if it docs not, it unites the parts furrounding the fuppurative centre, and confines the fuppuration to that point ; and as the abfcefs increafes in fize, the
adhefive inflammation fpreads uniting the parts as it fpreads,
fo that the general cavity is excluded.   Thus the fuppuration is confined to the firft point, and forms there a kind of
circumfcribed abfcefs, as will be more fully explained
hereafter.

In inflammations of the pleura or furface of the lungs,
the fame thing happens, for the adhefive inflammation

takes place, and the furfaces are united, which union going
before the fuppurative confines it to certain limits, fo that
diſtinct abſceſſes are formed in this union of the parts; and
the whole cavity of the thorax is not involved in a general
ſuppuration; ſuch caſes are called the ſpurious empyema.
The cellular membrane, every where in the body is united exactly in the ſame manner, the ſides of the cells throw
out, or as it were, ſweat out the uniting matter, which fills
the cavities and unites the whole into one maſs.

The adheſive inflammation often diſpoſes the parts to
form a cyſt, or bag; this is generally to cover ſome extraneous body that does not irritate ſo much as to produce
ſuppuration; ſuch as a ſack formed to incloſe a bullet,
pieces of glaſs, etc.

With the ſame wiſe views it unites the parts or cellular
membrane which lies between an abſceſs, and the ſpot
where that abſceſs has a tendency to open, as will be more
fully explained hereafter, when I come to treat of ulceration.

The lungs are ſo circumſtanced as to partake of the effects of two principles, the one as an internal uniting ſurface, the other as a ſecreting ſurface; the laſt of which
conſtitutes the peculiar ſtructure and uſe of this viſcus;
and the firſt is no more than the reticular or uniting ſubſtance of thoſe cells. The internal or the uniting membrane of the lungs unites readily by the adheſive inflammation, as in cellular membrane through the body generally;
but the air-cells, like the inner ſurface of the urethra, noſe,
inteſtines, etc. paſs directly into the ſuppurative inflammation, and therefore do not admit of the adheſive, by which
means the matter formed is obliged to be coughed up,
which produces ſymptoms peculiar to the parts affected;
and it is perhaps almoſt impoſſible to produce an inflammation on either of thoſe two ſurfaces without affecting the
other; which, probably is one reaſon why the treatment of
inflammation in thoſe parts is attended with ſuch bad ſuc-
ceſs.

We cannot give a better illuſtration of the uſe of the ad-
heſions produced in conſequence of this inflammation, than
to contraſt it with the eryſipelatous, of which I have already given an account.

When the eryfipelatous inflammation takes place, the
matter gets very freely into the furrounding and fo
cellular membrane, and then diffufes itfelf almoft over
whole body ; while, in another kind of conftitution, the
hefive inflammation would have been produced, to have
vented its progrefs.

A man was attacked with a violent inflammation on each
fide of the anus, which I did not fee till fome days after it
began.   It had the appearance of the fuppurative inflammation joined with the eryfipelatous ; for it was not fo circumfcribed as the fuppurative, nor did it fpread upon the
fkin like the true eryfipelatous, and the fkin had a fhining
œdematous appearance.   The inflammation went deeper
into the cellular membrane than the true eryfipelatous.
He was bled.   The blood was extremely fizy.   He took
a purge, and was fomented.   He had a difficulty in making water, moft probably from the preffure of the fwelling
upon the urethra.   The day following I obferved that the
fcrotum of that fide was very much fwelled, which had
extended up the right fpermatick chord ; on examining this
fwelling, I plainly felt a fluid in it, with air, which founded on being fhaken.   The cafe now plainly difcovered itfelf.   I immediately opened the fuppuration on each fide
of the anus, which difcharged a dufky coloured pus, very
fœtid, with a good deal of air.   Upon fqueezing the fwelling in the fcrotum, etc. I could eafily difcharge the matter
and air by thefe openings, therefore made him lie principally upon his back, and fqueeze thefe fwellings often,
with a view to difcharge this matter by the openings ; the
matter at the part where it was formed, was not contained
in a bag or abfcefs, but formed in the cellular membrane,
without previous adhefion.

The fcrotum now inflamed, and feemed to have a tendency
to open ; at leaft it looked livid and fpotted.   I opened it
at this part, and it difcharged a good deal of matter and air.
A general fuppuration came upon the whole cellular membrane of thofe parts, and the matter paffed up through the
cellular membrane of the belly ; and the cellular membrane
of the loins was loaded with matter, from its finking down
from the cells of the abdomen.   I made openings there,
and could fqueeze out a great deal of matter and air.   A
mortification came on juft above the right groin, and when

I removed the flough, matter was difcharged. I alfo made
openings on the loins, on the fide of the abdomen, etc. He
lived but a few days in this way, in which time the cellular
membrane was hanging out of the wounds like wet dirty tow.

The adhefive inflammation takes place in confequence
of accidents, when it is impoffible it fhould ever produce
the fame good effects, fuch as in wounds which are not allowed or cannot heal by the firft intention ; for inftance, a
ftump after amputation, and many other wounds; but it is one
of thofe fixed, and invariable principles of the animal machine
which upon all fuch irritations, uniformly produces the u-
niting procefs, though like many other principles in the fame
machine thefe effects, are perhaps not fo much required,
fo that although a wound is not allowed to heal, or cannot
heal by means of the adhefive inflammation, yet the furrounding parts go through the common confequences of
being wounded, and the furrounding cells are united, as
was defcribed when I treated of union by the firft intention.
It firft throws out the blood, as if the intention was to u-
nite the parts again ; the newly cut or torn ends of the
veffels, however, foon contract and clofe up, and then the
difcharge is not blood, but a ferum with the coagulating
part of the blood, fimilar to that which is produced by the
adhefive ftate of inflammation, fo that they go through the
two firft procefles of union ; therefore the ufe of the adhefive inflammation does not appear fo evidently in thefe cafes, as in fpontaneous inflammation ; however, in cafe of
wounds, which are allowed to fuppurate, it anfwers the
great purpofe of uniting the cells at the cut furface from
their being fimply in contact with each other, as has been
defcribed, which confines the inflammation to that part,
without which the irritation arifing from this ftate of imperfection might have been communicated from cell to cell,
and proceed farther than it commonly does. The cut veffels, by this means, are alfo united, which hinders the progrefs of inflammation from running along their cavities, as
we find fometimes to be the cafe in the veins of a wounded
furface, where this inflammation has not taken place. From
every thing which has been faid, it muft appear, that all furfaces which are fuppurating in confequence of this inflammation, have their bafis in that ftate of the adhefive inflammation,
which very nearly approaches to fuppuration ; and this inflammation is lefs and lefs, as it recedes further from the
fuppurating centre.

CHAPTER IV.

OF THE SUPPURATIVE INFLAMMATION.

———————————

WHEN the adhesive inflammation is not capable of resolution, and has gone as far as possible to prevent the necessity of suppuration, especially in those cases that might have admitted of a resolution, as in spontaneous* inflammations in general, where there has neither been an exposed laceration of the solids, nor, as beforementioned, loss of substance, but where the natural functions of the part have only been so deranged that it was unable to fall back into a natural and sound state again; or secondly, where it was a consequence of such accidents, as the effects of the adhesive could not in the least prevent, (as in wounds that were prevented from healing by the first or second intention) then under either of these two circumstances suppuration takes place.

The immediate effect of suppuration, is the produce of the pus, from the inflamed surface, which appears in such cases or under such circumstances to be a leading step to the formation of a new substance, called granulations, which granulations are the third method in the first order of parts, of restoring those parts to health ; but upon all internal canals, suppuration is certainly not a leading step to granulations, which will be explained hereafter.

The same theory of the adhesive inflammation respecting the vessels is, I believe, applicable to the suppurative ; for when suppuration is the first, we have the vessels in the same state as in the adhesive when it happens, but their dispositions and actions must have altered, there being a great difference in their effects.

---

* I have used this word to denote a case where no visible cause of inflammation existed ; for strictly there can be no such thing in nature as spontaneous.

This is so much the case, that the true inflammatory difposition and action almost immediately ceases upon the
commencement of suppuration ; and although the vessels
may be nearly in the same state, yet they are in a much
more quiescent state than before, and have acquired a new
mode of action.

I shall endeavour to establish, as an invariable fact, that
no suppuration takes place which is not preceded by inflammation ; that is, no pus is formed but in consequence
of it ; that it is an effect of inflammation only, is proved
in abscesses, from a breach in the solids attended with exposure, and from extraneous matter of all kinds, whether introduced or formed there. In abscess, suppuration is an
immediate consequence of inflammation ; from the exposure of internal cavities no suppuration comes on till inflammation has formed the disposition and action ; and although we find collections of extraneous matter, something
like pus, in different parts of the body, yet such
extraneous matter is not pus : however, towards the
last, in such collections, pus is often formed, but then
this is in consequence of inflammation having taken place
towards the surface, and when such collections are opened
they immediately inflame, universally similar to every branch
of the solids, and then the future discharge is pus, all of
which I shall now treat.

The irritation, which is immediately the cause of suppuration, is the same, from whatever cause it may proceed, similar to that which produces the adhesive stage ; it is a similar process going through the same stages, and is attended
with nearly the same circumstances, whether it takes it rise
from external violence, the constitution, or a disposition in
the part if all other circumstances are equal ; however, it
is not so general in its causes as the adhesive, for the thickening process will take place in many diseases, where true
suppuration is not admitted ; as in some scrofulous cases,
some venereal, and also cancers ; suppuration, therefore,
depends more on the soundness of parts than the adhesive,
and this is so much the case, that we can, in some degree,
judge of a sore simply by its discharge.

It appears very difficult to give a true and clear idea of
the whole chain of causes leading to suppuration. The immediate state of parts, which may be called the immediate
cause, I conceive to be such as cannot carry on its usual

functions of life, and which state of parts I have called the state of imperfection, let the cause of that state be what it will; we have shown that irritation simply is not always sufficient, it often only brings on the adhesive stage, which is in most cases intended to prevent the suppurative, as has been observed.

It is a curious fact, to see the same mode of action producing two such contrary effects, and each tending to a cure; the first producing from necessity the second, and being also subservient to it. Violence done to parts is one of the great causes of suppuration; but we have already remarked that violence simply does not always produce this inflammation: that it must be a violence followed by a prevention of the parts performing their cure in a more simple way, viz. a restoration of the structure, so as to carry on the animal functions of the part, or in other words, a prevention of union by the first or second intention, or attended with this circumstance of the parts being kept a sufficient time in that state into which they were put by the violence; or what is something similar to this, violence attended with death in a part, such as in many bruises, mortifications, sloughs in consequence of caustics, etc. which, when separated, have exposed internal surfaces.*

Various have been the opinions on this subject; and as every violence committed from without, under the circumstances before mentioned, is exposed more or less to the surrounding air, the applications of this matter to internal surfaces has generally been assigned as a cause of this inflammation; but air certainly has not the least effect upon those parts; for a stimulus would arise from a wound were it e-

---

* But here we may just remark, that the first processes towards suppuration in cases of mortification, where a separation must take place prior to suppuration, are different from the foregoing; because, the living surface is to separate from it the dead parts, and therefore another action of the living powers is required, which is what I call ulceration; and by the phænomena on this occasion, it would appear that nature can carry on two processes at one and the same time; for while the separation is carried on by the absorbents, the arteries are forming themselves for suppuration; so that at the same time the part is going through these two very different species of inflammation.

ven contained in a vacuum. Nor does the air get to the parts that form circumscribed abscesses so as to be a cause of their formation, and yet they as readily suppurate in consequence of inflammation as exposed surfaces.

Further, in many cases of the emphysema, where the air is diffused over the whole body, we have no such effect, and this air not the purest, excepting there is produced an exposure or imperfection of some internal surface for this air to make its escape by, and then this part inflames. Nay, as a stronger proof, and of the same kind with the former, that it is not the admission of air which makes parts fall into inflammation, we find that the cells in the soft parts of birds, and many of the cells and canals of the bones of the same tribe of animals, which communicate with the lungs*, and at all times have more or less air in them, never inflame ; but if these cells are exposed in an unnatural way, by being wounded, etc. then the stimulus of imperfection is given, and the cells inflame, and unite if allowed; but if prevented, they then suppurate, granulate, etc.   '

The same observation is applicable to a wound made into the cavity of the abdomen of a fowl ; for there the wound inflames and unites to the intestines to make it a perfect cavity again ; but if this union is not allowed to take place, then more or less of the abdomen will inflame and suppurate.

If it was necessary that air should be admitted in order for suppuration to take place, we should not very readily account for suppuration taking place in the nose from a cold, as this part is not more under the influence of air at one time than at another ; nor is the urethra in a gonorhœa affected by the air more at that time than at any other; these parts being at all times under the same circumstances with respect to air, therefore there must be another cause.

The sympathetic fever has been supposed a cause, which will be considered when I treat of the formation of pus.

In cases of violence I have endeavoured to give a tolerably distinct idea of the steps leading to suppuration ; but we are still at a loss with respect to the immediate cause of those suppurations which appear to arise spontaneously ; for in these it is almost impossible to deter-

* Vide Observations on certain Parts of the Animal Economy, page 89.

mine whether the inflammation itself be a real disease, viz,
an original morbid affection, or whether it may not be (as
is evidently the case from external violence) a salutary procefs of nature to restore parts whose functions, and perhaps texture, has been destroyed by some previous, and almost imperceptible disease or cause. Suppuration being,
in cases of violence, a means of restoration, affords a presump-
tion that it is a like instrument of nature in spontaneous
cases. If it is the first, viz. a real disease, then two causes
that are different in themselves can produce one effect, or
one mode of action, for the result of both is the same; but
if it is the last, then suppuration must be considered as depending on exactly the same stimulus being given, as in the
abovementioned instance of violence.

Suppuration does not arise from the violence of the action of the parts inflamed, for that circumstance simply rather tends to produce mortification; and we see that in the
gout which does not suppurate, there is often more violent
inflammation than in many others that do; all internal canals likewise suppurate with very flight inflammation, when
not in an irritable habit; but if of a very irritable disposition, the action will almost exceed suppuration, and by its
becoming milder, suppuration will come on.

But if we suppose the cause of inflammation to be a disposition in the parts for such actions, without the parts
themselves being either diseased, or in such state to be similar to the destruction or alteration of their texture, this inflammation then may arise from a vast variety of causes,
with which we are at present totally unacquainted; nay,
which we do not perhaps even suspect; and this last opinion, upon a flight view, would seem to be the most probable, because we can frequently put back these spontaneous
inflammations, which would not be the case if they came
on from the destruction of a part, or any thing else, whose
stimulus was similar to it; for no such thing can be done
with wounds, if they are not soon united by the first intention, they must suppurate; however, this argument is not
conclusive, for we can prevent suppuration in those arising
from accident, by uniting them by the second intention,
which is preventing suppuration, by acting as a kind of resolution.

Although suppuration is often produced without much
visible violence of action in the part, yet when it is a con-

sequence of a healthy inflammation, we find in general that the inflammation has been violent.

It is always more violent than in its preceding inflammation ; and in such cases it would appear to be little more than an increased action, out of which is produced an entire new mode of action, and which of course destroys the first.

It is from this violence that it produces its effects so quickly ; for the inflammation which is capable of producing quickly so great a change in the operations of the parts, as suppuration, must be violent ; because it is a violence committed upon the natural actions and structure of the parts.

This inflammation will also be more or less, according to the violence of the cause producing it, compared with the state of the constitution and parts affected.

The inflammation which precedes suppuration is much more violent in those cases where it appears to arise spontaneously, than when it arises from any injury done by violence. A suppuration equal in quantity to that from an amputation of the thigh, shall have been preceded by a much greater inflammation than that which is a consequence of the amputation.

This inflammation would seem to vary somewhat in its effects according to the exertion of that power during its progress ; for in proportion to its rapidity the cause is certainly more simple, and its termination and effects more speedy and salutary ; and this idea agrees perfectly with inflammation in consequence of accidents, for there it runs through its stages more rapidly, and with less inflammation ; necessity appears to be the leading cause here.

This seems to be the case even in those parts which have a tendency to slow and specific diseases ; as for example, in the breasts of women, or the testicles in men. For if these parts inflame quickly, the effects will be more salutary than if they inflamed slowly. In other words, those parts are capable of being affected by the common suppurative inflammation, which in most cases terminates well ; perhaps the specific inflammation is slow in its progress and operation, and such slowness marks it to be an inflammation of some specific kind.

In whatever light we view this fact, it at least leads us with more certainty to what the effects of an inflammation will be, and thus often to form a just prognostic.

Suppuration takes place much more readily in internal canals, than internal cavities.

Suppuration takes place more readily upon the surface of canals than in either the cellular or inverting membrane. The same cause which would produce a suppuration in the first parts, would only produce the adhesive in the other; for instance, if a bougie is introduced into the urethra, for a few hours, it will produce suppuration; while, if a bougie was introduced into either the tunica vaginalis testis, or the abdomen, but for a few hours, it would only give the disposition for adhesions, and even might not go the length of this stage of inflammation in so short a time; but such surfaces often produce a greater variety of matter than a fore, it is not always pus; and this, probably, arises from the cause not being so easily got rid of. An irritation in the bladder from stone, stricture in the urethra, or disease of the bladder itself, gives us a great variety of matter; pus, mucus, slime, are often all found; sometimes only one or two of them. I have some idea that the mucus is easiest of production; but I am certain, that for the formation of slime, the greatest irritation is required.

I. THE SYMPTOMS OF THE SUPPURATIVE INFLAMMATION.

This inflammation has symptoms common to inflammation in general; but it has these in a greater degree than the inflammation leading to it, and has also some symptoms peculiar to itself; it therefore becomes necessary to be particular in our description of these peculiarities.

The sensations arising from a disease generally convey some idea of its nature; the suppurative inflammation gives us as much as possible the idea of simple pain, without having a relation to any other mode of sensation: we cannot annex an epithet to it, but it will vary in some degree, according to the nature of the parts going into suppuration, and what was remarked, when treating of the adhesive state, is in some degree applicable here.

This pain is increafed at the time of the dilating of the arteries, which gives the fenfation called throbbing, in which every one can count his own pulfe from paying attention merely to the inflamed part ; and perhaps this laft fymptom is one of the beft characteriftics of this fpecies of inflammation. When the inflammation is moving from the adhefive ftate to the fuppurative, the pain is confiderably increafed, (and which would feem to be the extent of this operation in the part) ; but when fuppuration has taken place, the pain in fome degree fubfides ; however, as ulceration begins, it in fome degree keeps up the pain, and this is more or lefs, according to the quicknefs of ulceration, but the fenfation attending ulceration gives more the idea of forenefs.

The rednefs that took place in the adhefive ftage is now increafed, and is of a pale fcarlet. This is the true arterial colour, and is to be accounted a conftant fymptom, as we find it in all internal inflammations, when at any time expofed, as well as in thofe that are external.

Befides, I obferved in the introduction to inflammation, and when treating of the adhefive ftate, that the old veffels were dilated, and new ones were formed ; thefe effects, therefore, are here carried ftill farther in the furrounding parts, which do not fuppurate, and conftitute two other caufes of this rednefs being increafed, by the veffels becoming ftill more numerous, and the red part of the blood being pufhed more forward into many veffels, where only the ferum and coagulating lymph went before.

The part which was firm, hard, and fwelled, while in the firft ftages, or the adhefive ftate, now becomes ftill more fwelled by the greater dilatation of the veffels, and greater quantity of the extravafated coagulating lymph thrown out, in order to fecure the adhefions.

The œdematous fwelling furrounding the adhefive gradually fpreads into the neighbouring parts.

In fpontaneous fuppurations one, two, three, or more parts of the inflammation lofe the power of refolution, and affume exactly the fame difpofition with thofe of an expofed furface, or a furface in contact with an extraneous body. If it is in the cellular membrane that this difpofition takes place, or in the invefting membranes of circumfcribed cavities, their veffels now begin to alter their difpofition and mode of action, and continue changing till they gradually form themselves to that state which fits them to form pus ; so that the effect or discharge is gradually changing from coagulating lymph to pus; hence we commonly find in abscesses, both coagulating lymph and pus, and the earlier they are opened the greater is the proportion of the former. This gave rise to the common idea and expression, " That the matter is not concected ;" or, " The abscess is not yet ripe." The real meaning of which is, the abscess is not yet arrived at the true suppurative state.

From hence it must appear that suppuration takes place upon those surfaces without a breach of solids or dissolution of parts, a circumstance not commonly allowed * ; and when got beyond the adhesive state they become similar in their suppuration to the inner surfaces of internal canals.

There is a certain period in the inflammation, when the suppurative disposition takes place, which is discovered by new symptoms taking place in the constitution, viz. the shivering.

Although the sudden effects produced in the constitution would shew that this change of disposition is pretty quick, yet its effects in the parts must be far from immediate, for some time is required for the vessels to be formed by it, so as to produce all the consequences intended by nature ; and, indeed, we find it is some time before suppuration completely takes place ; and that it is sooner or later according as the inflamed state is backward in going off; for while the inflammation lasts the part, as it were, hangs between inflammation and suppuration.

The effect of inflammation appears to be the producing of the suppurative disposition, or that state of a part which disposes it to form pus ; in doing this the inflammation

* The knowledge of this fact in some of the larger cavities is not quite new ; for I remember, about the year 1749 or 1750, that a young subject came under our inspection, and on opening the thorax it was found on the left side to contain a considerable quantity of pus. Upon examining the pleura and surface of the lungs, they were found to be perfectly entire. This was taken notice of by Dr. Hunter as a new fact, that suppuration could take place without a breach of surface ; and he sent to Mr. Samuel Sharp to see it. It was also new to him, and he published it in his Critical Inquiry. Since that period it has been often observed in the peritoneal inflammation,

seems first to be carried to such a height as to destroy that
state of the parts on which itself depends, the consequence
of which is, that they lose the inflammatory disposition,
and come into that which fits them for forming pus.

It seems to be a fixed and most useful law in the animal
œconomy, that in spontaneous inflammation, when it has
either destroyed the natural functions of parts, so much as
to prevent their returning by a retrograde motion, as it
were, to the state from whence they set out, or where the
first cause was a destruction of the natural functions, as an
exposure of internal surfaces, that they form a disposition
to the second method of cure. That the disposition for
suppuration is very different from the actual state of inflammation, though produced by it, is proved by a variety
of observations; for no perfect suppuration takes place till
the inflammation is gone off; and as the inflammation
ceases, the disposition to suppuration gradually comes on.
If too by any peculiarity in the constitution or inflammation by which it is continued, or if by any accident an inflammation is brought on a healthy sore, the discharge and
other appearances become the same as they were when
the part from whence they arose was first in an inflamed
state, very different from those observed when it was arrived at the state of a kind of suppuration.

II. OF THE TREATMENT NECESSARY IN IN-FLAMMATION, WHEN SUPPURATION MUST TAKE PLACE.

In cases of inflammation arising from accident, but so
circumstanced, that we know suppuration cannot be prevented, the practice will be to moderate the inflammation,
if necessary, but not with a view to prevent suppuration;
for if the powers are very great, and the violence committed
very considerable, the inflammation will probably be very
violent; and if it should have equal effects on the constitution, which will be in proportion to the quantity of surface inflamed, then certain constitutional means of relief will be necessary, such as bleeding, purging, regi-

men, and perhaps producing sickness ; because while that inflammation continues to have effects upon the constitution, the suppuration which takes place will not be so kindly, as it would otherwise be ; but if the constitution is of the irritable kind, which will be generally known by the inflammation, the same practice as mentioned above is neceffary ; in short, whatever is to be the confequence, whether refolution or fuppuration, the irritability and the too great action of the veffels, whether arifing from too great powers, or too great action with fmall powers, are to be corrected or removed, as they in all cafes counteract falutary operations.

In cafes where the conftitution has fympathized much with the inflamed part, fuch medicines as produce a flight perfpiration much relieve the patient ; fuch as antimonials, Dover's powders, faline draughts, fpirits of mendereries, etc. becaufe they endeavour to keep up an univerfal harmony, by putting the fkin in good humour, which quiets every fympathizing part, and by counteracting the effects of the irritability. Opium often leffens actions, although it feldom alters them, when only given as an opiate, and may be of a temporary fervice : however, this is not always a confequence of opium, as there are conftitutions where it increafes irritation, and of courfe difeafed action.

Frefh wounds, confidered fimply as wounds, are all of the fame nature, and require one uniform treatment; the intention being to put them into that fituation in which they can fuppurate with moft eafe to themfelves ; and the firft dreffings commonly remain till fuppuration comes on, unlefs fome peculiarity from the fituation of parts, or other collateral circumftances, fhould make it neceffary to remove the dreffings or vary the treatment.

The difference between one wound and another, with refpect to the nature of the part wounded, will vary very much ; fome will have fmall veffels wounded that cannot be conveniently got at in order to tie them up, yet fhould be ftopt from bleeding, which can be done by the mode of dreffing, and therefore require dreffing fuitable to this circumftance alone.

Wounds opening into cavities where peculiarities of the contained parts are joined with the injury done to them by

Vol. II. O

the accident will require a suitable mode of dressing ; the influence too that a simple wound in the containing parts may have upon those in the cavity, as a wound into the belly, thorax, joints, skull, etc. will oblige the surgeon to vary the mode of dressing from that of a simple wound. While many wounds will require being kept open for fear of uniting a-gain, in order to answer some future purpose, as the wound made into the tunica vaginalis testis, for the radical cure of the hydrocele ; others may require attention being paid to them before suppuration comes on, and therefore should be so dressed as to admit of being soon and easily removed, to examine the parts occasionally as the symptoms arise. This ought to be the case in wounds of the head, attended or not attended with fracture of the skull.   But whatever mode of application may be thought necessary to answer the various attending circumstances, yet as they are all wounds which are to come to suppuration, one general me-thod is to be followed respecting them all, as far as those peculiarities will allow.

The application which has been made to wounds for some years past in this country, has been in general dry lint ; what brought this application into common practice, most probably was, its assisting in stopping the hemorrhage ; and as most wounds are attended with bleeding, it became universal ; but as it became universal, it lost the first in-tention, and became simply a first dressing.

I need hardly remark here, that all wounds that are to suppurate are first attended with inflammation, and there-fore are so far similar to spontaneous inflammations which are to suppurate. If this observation is just, how contradic-tory must this mode of treatment be to common practice, when spontaneous inflammation has already taken place ; for let me ask, where is the difference between an inflam-mation with a wound, and one without ? And also, what should be the difference in the application to a part that is to inflame, (while that application is made to the part) and one applied to an inflammation which has already taken place ? The answer I should make to such a question is ; there is no difference.

Wounds that are to suppurate, I have already observed, are first to go through the adhesive and suppurative inflam-mation.   These inflammations in a wound, are exactly si-milar to those spontaneous inflammations which suppurate

and form an abscess, or those inflammations which ulcerate on the surface, and form a sore.

The applications to these which are now in practice, I
have formerly observed, are poultices and fomentations ;
these, however, appear to be applied without much critical
exactness or discrimination, for they are applied before suppuration has taken place, and where it is not intended it
should take place; they are applied to inflammations where
it is wished they should suppurate ; and applied after suppuration has taken place. Now, with respect to suppuration
itself, abstracted from all other considerations, the indication
cannot be the same in all of those states ; but if poultices
and fomentations are found to be of real service in those two
stages of the disease, then there must be something common
to both, for which they are of service, abstracted from simple suppuration. I also formerly observed, that poultices
were of service when the inflammation had attacked the
skin, either of itself, or when an abscess had approached so
near that the skin had become inflamed, and that this service consisted in keeping the skin soft and moist. This appears to m to be the use of a poultice in an inflammation,
either before suppuration or after, as inflammation still exists, till it is opened ; for inflammation is necessary in an
abscess, while it is making its approaches to the skin, which
I have called the ulcerative, and then, and only then, it begins to subside ; it is therefore still proper, in as much as it
is of service to inflammation ; so far, therefore, their practice is right and consistent, as the first reason exists through
the whole ; but when applied to inflamed parts, which are
meant not to suppurate, their reasoning or principles upon
which they applied it must fail them, although the application is still very proper.

If my first proposition is just, viz. that wounds which are
allowed to suppurate, are similar to inflammations that are
also to suppurate ; then let us see how far the two practices
agree with this proposition. Lint, I have observed, is applied to a fresh wound, which is to inflame ; and the same
lint is continued through the whole of the inflammation till
suppuration comes on, because it cannot be removed. Lint,
considered simply as an application to fresh wounds which
are to inflame, is a very bad one, for it more or less adheres
to the surface of the wound, by means of the extravasated
blood ; hence it becomes difficult of removal, and often

shall remain in fores for months, being interwoven with the granulations, especially when applied to the surface of circumscribed cavities, such as the tunica vaginalis testis, after the operation for the hydrocele ; however, this is not always the greatest inconvenience, the circumstance of its being loaded or soaked with blood, subjects it to become extremely hard when it dries, which it always does before the separation takes place, which separation is only effected by the suppuration.   In this way it becomes the worst dressing possible for wounds.

As poultices are allowed by most to be the best application to an inflamed part, not attended by, or a consequence of a wound, but considering it simply as an inflammation, I do conceive that the same application is the best for every inflammation, let it be from whatever cause ; for the idea I have of the best dressing to a wound, simply as a wound which is to inflame, is something that keeps soft and moist, and has no continuity of parts, so that it is easily separated.   The only application of this kind is a poultice, which, from these qualities, is the very best application to a fresh wound.   It keeps it soft and moist, and is at all times easily removed, either in part or the whole.

The same medical advantage is gained here, as when it is applied to an inflamed part ; but although it had not these advantages, yet the circumstance of being easily removed is much in its favour, especially when compared to dry lint.

But a poultice from other circumstances, cannot at all times and in all places be conveniently applied.

To preserve the above properties, it is necessary there should be a mass, much too large for many purposes ; but when they can be used with tolerable convenience, they are the best applications.   When they cannot be applied with case, I should still object to dry lint, and would therefore recommend the lint to be covered with some oily substance, so that the blood shall not entangle itself with the lint, but may lie soft, and come easily off.

This mode of dressing should be continued for several days, or at least till fair suppuration comes on, and when that has taken place, then dry lint may be with great propriety used, except the fore is of some specific kind, which is seldom the case in fresh wounds ; for accidental wounds seldom happen to specific diseases ; and a wound in confequence of an operation fhould not be fpecific, becaufe the fpecific affection (if there is any) fhould have been removed by the operation ; and fhould therefore be a wound in the found part ; as after an amputation of a fcrofulous joint, or the extirpation of a cancerous breaft.   Or if they take on fome fpecific difpofition afterwards, then they muft be dreffed accordingly, as will be explained hereafter.

Poultices are commonly made too thin, by which means the leaft preffure, or their own gravity, removes them from the wound ; they fhould be thick enough to fupport a certain form when applied.

They are generally made of ftale bread and milk ; this compofition, in general, makes a too brittle application, it breaks eafily into different portions, from the leaft motion, and often leaves fome part of the wound uncovered, which is fruftrating the firft intention.

The poultice which makes the beft application, and continues moft nearly the fame between each dreffing, is that formed of the meal of linfeed ; it is made at once‡ and when applied it keeps always in one mafs.

Fomentations are generally applied at this ftage of the wound, and they generally give eafe at the time of application, which has (joined with cuftom) been always a fufficient inducement to continue them.   As foon as fuppuration is well eftablifhed, the part may then be dreffed according to the appearaces of the fore itfelf.

The kind of wound to which the above application is beft adapted, is a wound made in a found part, which we intend fhall heal by granulation.   The fame application is equally proper, where parts are deprived of life, and confequently will flough.   It is therefore the very beft dreffing for a gun-fhot wound, and probably for moft lacerated wounds.   For lint applied to a part that is to throw off a a flough, will often be retained till that flough is feparated, which will be for eight, ten, or more days.

‡ Take boiling water. q. s. and ftir in the linfeed till it becomes of a fufficient thicknefs, and then add a fmall quantity of fome fweet oil.

In the treatment of wounds that are to suppurate, it is in one view of the subject right to allow the parts to take their natural and spontaneous bent. From the natural elasticity of the skin, and the contraction of muscles, the parts wounded are generally exposed, and from the consequent inflammation, they generally become more so. This is commonly more the case in wounds produced by accident ; for as a small wound and much old skin are always desirable, surgeons very wisely are anxious to wish for both. In many operations, they are desirous of preserving the skin, viz. where they are removing parts, as a limb ; dissecting out tumors, or opening an abscess ; all of which is extremely proper, and they continue to practice upon this princciple immediately upon the receiving the wound, and in performing any of the abovementioned operations ; for the skin, after amputation, is drawn down, and bound down, and the wounds are pressed together by bandages. In one point of view, this is beginning too early ; it is beginning it when nature has the very opposite principle in view. Inflammation, the parts must submit to ; and as inflammation by its effects will generally have a tendency to make them recede more, in this light it is proper not to check the effects of inflammation, therefore let them take their own way till inflammation subsides, and granulations are formed, which granulations, I have already observed, by their power of contraction, will do what we wanted to have done ; and if, from some of the first circumstances not being properly attended to, the contraction of the granulations is not sufficient, then is the time to assist, and not before. However, if we take up this in another point of view, we shall see a considerable utility arising from bringing the skin as much as possible over the wound, and keeping it there ; for in the time of inflammation the parts will adhere or unite in this situation, by which means the sore will be less than it otherwise would; and I conceive that this practice, when begun, should be for some time continued, for fear the adhesions may not be sufficient to stand their ground till the granulations can assist.

It often happens in many wounds, both from accident and operations, that part of the wound may with great propriety be healed by the first intention ; such as in many accidents on the head, when a part of the scalp has been torn off, on the face, etc. as also after many operations, especially where the skin is loose, as in the scrotum ; or where the skin has been attended to in the time of the operation, as in some methods of amputation, extirpation of breasts, etc. a part of the saved skin, etc. may be made to unite to the parts underneath by the first intention, and therefore only part of the wound allowed to suppurate; in all such cases, a proper contracting or sustaining bandage may be applied with great advantage; even stitches may be used with great propriety, as was recommended in the healing wounds by the first intention.

III. THE TREATMENT OF THE INFLAMMATION WHEN SUPPURATION HAS TAKEN PLACE

In spontaneous inflammations, whether from a constitutional or local affection, when suppuration has taken place, it is most probable that another mode of practice must be followed than that which was pursued to prevent it; but even now, if a stop could be put to the further formation of matter, after it has begun, it would in many cases be very proper, and still prevent a great deal of mischief. Suppuration does certainly sometimes stop, after having begun, which shows that there is a principle in the animal œconomy of diseases from which the machine is capable of producing this effect[*].

[*] I have formerly observed, that the inflammation goes off often without producing suppuration: and I have also mentioned instances of suppuration going off without the parts having produced granulations, and then the parts fall back into the adhesive state, and the matter being absorbed, they are left in nearly the same state as before the inflammation came on, as a presumptive proof of this, in many of the large cavities, which have been allowed to inflame and suppurate, (by having been opened) we find them often doing well without ever forming granulations, and that suppuration generally goes off; and I do not believe ever fall back into the adhesive state, so as to unite the parts, but the parts resume their original and natural state or disposition, and no adhesions are formed; this appears sometimes to happen in cases of the empyema after the operation has been performed; I have seen cases where wounds had been made into the cavity of the thorax, where there was every reason to suppose the whole cavity was in a state of suppuration, and yet those patients got well. I can hardly suppose that in these cases the parts had granulated and united in the cure, as the cellular

I have feen buboes cured by vomits, after fuppuration
had been confiderably advanced ; and it is a very common
termination of fcrofulous abfceffes ; but in fcrofulous abfceffes we very feldom find inflammation ; this procefs appears to be a leading circumftance in ulceration, which is
the very reverfe of union ; even in fuperficial fores, which
are the moft likely to continue fuppuration, if excited, we
find by allowing them to fcab, when they will admit of it,
that the act which admits of fcabbing is the very reverfe of fuppuration, and it ceafes ; however, it is a procefs which the animal œconomy does not readily accept of,
and our powers in producing this effect are but very fmall ;
if thefe powers could be increafed by any means, it would
be a falutary difcovery ; becaufe fuppuration itfelf, in many cafes, proves fatal ; for inftance, fuppuration of the brain
and its membranes; of the thorax and its contents ; as well
as of the abdomen and its contents ; in fhort, fuppuration
of any of the vital parts often kills of itfelf, fimply from the
matter being produced ; but this practice will by moft be
forbid in many cafes of fuppuration ; for it is fuppofed this
very fuppuration is a depofit of matter or humors already
formed in the conftitution ; but it is to be hoped that time
and experience will get rid of fuch prejudices.

When fuppuration cannot be ftopped or refolved, then
in moft cafes it is to be hurried on, which generally is the
firft ftep taken by furgeons.

membrane does ; becaufe I have feen many fimilar cafes, where
the patients have died, and no granulations have been found ;
and I have feen cafes of the hydrocele attempted to be cured radically by the cauftic ; when the flough came out, fuppuration came on ; but the orifice healing too foon, fuppuration has
ceafed, and the cure was thought to be completed ; but a return of the difeafe has led to another attempt and by laying open
the whole fack, it has been found that the tunica vaginalis was
perfectly entire. In fuch the fluids were a mothery ferum. I
have feen abfceffes go back in the fame manner : but I believe
that this procefs is more common to fcrofulous fuppurations than
any other ; and I believe to the eryfipelatous. I have feen joints
heal after having fuppurated and been opened, without having
produced granulations leaving a kind of joint. even when the
cartilages have exfoliated from the ends of the bones, which was
known by the grating of the two ends of the bones on one another.

How far suppuration can be increased by medicine or application I do not know ; but attempts are generally made ; and thence we have suppurating cataplasms, plasters, etc. recommended to us, which are composed of the warmer gums, seeds, etc. but I doubt very much if they have considerable effect in this way ; for if the same applications were made to a sore, they would hardly increase the discharge of that sore, probably rather decrease it ; however, in many cases, where the parts are indolent, and hardly admit of true inflammation, in consequence of which a perfect suppuration cannot take place· by stimulating the skin, a more salutary inflammation may be produced, and of course a quicker suppuration : but in the true suppuration, where inflammation preceded it, I believe it is hardly necessary to do any thing with respect to suppuration itself ; however, from experience, I believe these applications have been found to bring the matter faster to the skin, even in the most rapid suppurations, which was supposed to be an increased formation of matter ; but it can only be in those cases where the inner surface of the abscess is within the influence of the skin. This effect arises from another cause or mode of action being produced, than that of quickening suppuration, which is the hastening on of ulceration. I have mentioned that ulceration was an effect of, or at least attended by inflammation ; and, therefore, whatever increases that inflammation, will also increase the ulceration, which will bring the matter sooner to the skin, without an increased formation of matter.

Poultices of bread and milk are commonly used to inflamed parts when suppuration is known to have taken place ; this application can have no effect upon suppuration, excepting by lessening inflammation, or rather making the skin easy under it ; for we observed, that true suppuration did not begin till inflammation was abated ; but the inflammation must have reached the skin before poultices can have much effect, for it can only affect that part.

It may be thought necessary that the case of the patient should be considered, and we find that fomentations and poultices often produce that effect ; we find too, that by keeping the cuticle moist and warm, the sensative operations of the nerves of the parts are soothed, or lulled to rest ; while, on the contrary, if the inflamed skin is allow-

ed to dry, the inflammation is increased, and as probably
suppuration is not checked by such treatment, it ought to
be put into practice ; as warmth excites action, it is probable, the warmer the fomentation, so much the better ; and
in many cases the action is increased so that they can hardly bear it.

IV. COLLECTIONS OF MATTER WITHOUT INFLAMMATION.

I HAVE hitherto been describing true suppuration, which
I have said, " I believe is a consequence only of inflammation," a process generally allowed. Also in treating on
the cause of suppuration, viz. inflammation, I hinted, that
there were often swellings, or thickening of parts without
the visible or common symptoms of inflammation, viz.
without pain, change of colour, etc. and I also hinted, in
treating of suppuration, that there were collections of matter somewhat similar to suppuration, which did not arise
in consequence of the common inflammation ; these I shall
now consider : I conceive all such collections of matter to
be of a scrofulous nature ; they are most common in the
young subject, and seldom found in the full-grown or old.
It is commonly called matter, or pus, and therefore I
choose to contrast true suppuration with it.   Although I
have termed this suppuration, yet it has none of its true
characters, any more than the swellings, which are the
forerunners of it, have the true characters of inflammation ;
and as I did not call them inflammatory, strictly speaking, I
should not call this suppuration ; but I have no other term
expressive of it.

Many indolent tumors, slow swellings in joints, swellings of the lymphatic glands, tubercles in the lungs, and
swellings in many parts of the body, are diseased thicken-
ings, without visible inflammation ; and the contents of
some kinds of incysted tumor ; the matter of many scrofulous suppurations, as in the lymphatic glands ; the suppuration of many joints, viz. those scrofulous suppurations in
the joints of the foot and hand ; in the knee, called white

swellings ; the joint of the thigh, commonly called hip cafes ; the loins called lumbar abscesses ; the discharge of the abovementioned tubercles in the lungs, as well as in many other parts of the body, are all matter formed without any previous visible inflammation, and are therefore, in this one respect, all very similar to one another. They come on insensibly ; the first symptom being commonly the swellings, in consequence of the thickening, which is not the case with inflammation, for there the sensation is the first symptom.

These formations of matter, although they do approach the skin, yet do not do it in the same manner as collections of pus. They do not produce readily either the elongating or the ulcerative process, and as the matter was not preceded by the adhesive inflammation, these collections are more easily moved from their original seat into some other part, by any slight pressure, such as the weight of their own matter, which ·I have called abscesses in a part, in opposition to abscesses of a part : when the matter does approach the skin, it is commonly by merely a distention of the part, coming by a broad surface, not attended with any marks of pointing.

Their surrounding parts or boundaries are soft, not being attended with thickening ; more especially those in a part.

Such collections of matter are always larger than they would have been, if they had been either a consequence of inflammation, or attended by it ; this is owing to their indolence, allowing of great distention beyond the extent of the first disease, even moving into other parts, whereas an abscess, in consequence of inflammation, is confined to the extent of inflammation that takes on suppuration, and its rapid progress towards the skin prevents distention, and of course extention of the disease.

All those formations of matter, not preceded by inflammation, nor a consequence of it, are, I believe, similar to each other, having in this respect one common principle, very different from inflammation. The cancer although it produces a secretion, yet does not produce pus till exposed ; it is, therefore, one of those diseases, like the scrofula, which does not suppurate till inflammation comes on, and even seldom then ; for true suppuration arises from inflammation, terminating in a disposition to heal, which is not the case with cancer. In the scrofulous suppuration there is often a like reluctance to heal.

The kind of matter is another distinguishing mark, between that produced in consequence of inflammation, and what is formed without it; the last being generally composed of a curdly substance, mixed with a flaky matter, the curdly substance is, we may suppose, the coagulating lymph deprived of its serum *, and the other, or flaky, is probably the same, only in smaller parts; it looks like the precipitate of animal matter, from an acid or alkali.

So far these productions of matter in their remote and immediate cause, are not in the least similar to that arising from common inflammation, nor is the effect, viz. the matter similar; and to show still further, that suppuration is always preceded by inflammation, the very surfaces which formed the above matter, immediately produced true matter, when the inflammation comes on, which it always does whenever opened; which I shall now consider.

Since they are not similar in their causes or modes of production, let us next examine how far they are similar in their first steps towards a cure.

All parts which form matter of any kind, viz. whether in consequence of inflammation or otherwise, must go through similar processes to produce the ultimate effect or cure : the first step in either, is the evacuation of this matter, for till this is effected, nature cannot pursue the proper means towards a cure ; and if opened, the second step is granulation, and the third cicatrization. To accomplish the evacuation of the matter, there are two modes, one is the absorption of the matter, which is very common in the scrofula, or those productions of matter, not preceded by inflammation. This produces no alteration in the part, except that it gradually creeps into a sound state, the parts uniting again that had been separated by the accumulation of the matter ; it produces, also, no alteration in the constitution. Absorption, however, seldom takes place in suppuration, which is the consequence of inflammation. The other mode of discharging this matter is either by opening

* I may observe here, that the coagulating lymph of long standing, is not similar to the recent. This is similar to blood in general, for we find that the blood in aneurisms, which was first coagulated, is very different from that which has only coagulated lately.

the abfcefs, in order to allow it to pafs out, or by allowing ulceration to take place from the infide to produce its e-fcape ; and this procefs, in the prefent cafe, having peculiarities different from thofe arifing from inflammation, it is neeeffary they fhould be underftood. Ulceration, in confequence of fuppuration arifing from inflammation, is very rapid, efpecially if the fuppuration is fo likewife ; but ulceration, in confequence of matter being formed, which is not the effect of inflammation, is extremely flow ; it will remain months, even years, before the parts have completely given way ; they commonly come to the fkin by a broad furface, and not pointing like a circumfcribed abfcefs in confequence of inflammation ; fo far are thefe two different.

V. OF THE EFFECTS SUCH FORMATIONS OF MATTER HAVE ON THE CONSTITUTION.

WHATEVER may be the extent of fuch collections of matter, they feldom or ever affect the conftitution, unlefs they are feated in a vital part, or fo connected with it as to difturb its functions.

This is an effect of indolence in any difeafe. A young perfon fhall have a lumbar abfcefs, for inftance, for years, without a fingle conftitutional fymptom. It fhall appear to be making its way through a number of parts, fuch as the loins behind, the buttocks, the lower part of the abdomen before, and through the upper part of the thigh ; and in each part fhall fhew large collections of matter. All thefe fhall even attend the fame perfon, yet not any bad fymptoms, no fhiverings fhall accompany this fuppuration.* In fome there is not even the leaft degree of lamenefs, but this is often the firft ftage of the difeafe in the lumbar abfcefs.

---

* I have heard furgeons afk fuch patients, if they had rigors, even alluding to the time of increafe ; this was applying the idea of the fymptom of one difeafe to another, and alfo the firft ftage of a difeafe to the fecond.

Let us next confider and compare the confequences attending thefe two collections of matter when opened. When an abfcefs, in confequence of inflammation is opened, it immediately proceeds towards a cure, and perhaps it may have gone fome fteps towards a cure before opening, the inflammation ftill leffens, the fuppuration becomes more perfect, granulations begin to form, and all of thefe fteps naturally take place, becaufe inflammation had been the caufe ; but when a collection of matter, not preceded by inflammation, is opened, a very different procefs is firft to take place, viz. inflammation is now excited over the whole cavity of the abfcefs, which afterwards produces a perfect matter, fimilar to that produced in confequence of inflammation, when it is the original difeafe ; and which now produces its conftitutional affection, if it is fuch as to have connexion with the conftitution; but this will depend on the fize of the abfcefs, the fituation, and the nature of the parts, etc. however, it fometimes happens that they inflame before they are opened ; but this is in confequence of the matter diftending the cavity, and thereby acting as an extraneous body. I have feen white fwellings in the knee inflame before they were opened, then ulceration takes place, and the pus brought foon to the fkin, even after it had been confined for months, without producing the leaft tendency to ulceration, becaufe there had been none to inflammation; but the confinement of the matter becomes a caufe of the inflammation, and then ulceration takes place.

The inflammation and new fuppuration taking place in confequence of opening into thefe abfceffes is exactly fimilar to thofe arifing in confequence of wounds or openings made into natural cavities ; it was ftill, therefore, neceffary, that they fhould go through all the common fteps towards reftoration ; but, unfortunately, fuch inflammations have begun at the wrong end ; they have alfo fet down upon a fpecific difeafe, which they can feldom alter to their own nature. The inflammation is in fuch cafes extended over a much larger furface than the original ; which is not the cafe in abfcefs, in confequence of inflammation, for there the inflammation was the caufe and confined to the point.

In fome cafes, as in lumbar abfcefs, the extent of furface to inflame is immenfe, in comparifon to the extent of the original difeafe, and of courfe, when fuch abfceffes inflame, the symptoms in the constitution are in the same proportion.

How different is this from the opening of the abscefs in consequence of inflammation! There we have no inflammation following, except, what arifes in consequence of the wound made in the solids in the operation of opening; but when it is allowed to open of itself, there is no consequent inflammation, but suppuration goes on. But it would appear that when those collections of matter are allowed to open of themselves, that the succeeding inflammation does not so readily take place, as when opened by art. I have seen large lumbar abscesses open of themfelves on the lower part of the loins, which have difcharged a large quantity of matter; then clofed up, then broke out anew, and so on for months, without giving any other disturbance; but when opened, so as to give a free difcharge to the matter, inflammation has immediately fucceeded, fever has come on, and from the fituation of the parts inflamed, as well as their extent, death in a very few days after has been the confequence; it, therefore, often becomes a queftion whether we should inlarge the firft opening or not. We may obferve in general, that in cafes of this kind, where they are to terminate ill, that is, where they cannot be cured, and are fuch as to affect the conftitution, the confequent inflammation upon opening them, which produces the fympathetic fever, has that fever commonly terminating in the hectic, or continued into the hectic before any recefs takes place, so that the one is continued into the other, without any intermiffion; however, this is not always the cafe, and thofe variations will depend on the ftate of the fore, the ftate of the conftitution, etc.

VI. THE EFFECTS OF THE SUPPURATIVE IN-FLAMMATION ON THE CONSTITUTION.

It is to be obferved, that every local complaint of any confequence, or which has confiderable and quick action within itfelf, although not of confiderable magnitude, affects more or lefs the conftitution, and gives rife to what has been called the fymptomatic fever. Thefe fymptoms are

the sympathies of the constitution with a local disease or injury, and will vary according to a vast variety of cicumstances.
They will vary according to the nature of the constitution-
which admits of great differences, and which will include
different ages ; they will vary according to the nature of the part in a state of disease, which also admits of great differences ; they will vary according
to the quantity of mischief done, as well as the manner of its being done : that is, whether so as to call forth
immediate inflammation as a wound ; or, not so immediate,
as from having only killed a part ; they will vary according
to the situation of similar parts in the body ; and they will
vary according to the stage of the disease. This last variation may be divided into two kinds, the one which begins
slowly and increases progressively, as in the venereal disease, and the sympathetic affections of course come on gradually ; the other, where it begins at once with violence
and diminishes. The first of this last division we have nothing to do with at present ; it is, therefore, the kind of
constitution, the kind of parts, the diseases which commence with so much violence as to affect the constitution
at once with the constitutional effects arising from the local disease being incurable, that form our present subject.
I shall observe here, that every disease, whether local or
constitutional, that has the power of termination in itself,
commonly has its regular progress and stated times of action ; in some, however, there are no changes in the modes
of action, the disease coming on and dying away ; but in
others there are ; and in those where changes take place, there
are stated periods for those changes, so as to render them regular.   As regularity in the modes of action in disease is
conducive to the termination of that disease, it is a thing
very much desired ; for these changes are a cessation of
the action, either temporary or permanent.   As the constitution sympathises with a local irritation, and as that sympathy is according to the constitution, to the violence of
the irritation, and to the nature of the parts irritated ; and
the symptoms of that sympathy must be similar to constitutional complaints that are commonly taking place ; and if
the local complaints should not be known, then they will be
taken for constitutional complaints wholly, and treated as
such ; but often from their continuance, some local affection is suspected ; local complaints, however, are common-

ly preceded or attended with some local symptom either
directly or indirectly, or with some collateral symptom or
symptoms, so as to direct us to the cause. Local complaints
attended with inflammations, the objects of surgery, are
often attended with, or rather consequent upon violence of
some kind; such as the loss of a part, either fluid or solid,
which the constitution feels, and which loss, or violence,
adds to the constitutional affection. This will be according
to the quantity of injury or loss of living matter, whether
blood, or some solid, the time in the operation, the state of
the parts operated upon, and the nature of the part removed. I have seen a man die almost immediately upon the
loss of a testicle. I have seen convulsions immediately attend the operation for the hydrocele, so that I have almost
despaired of recovery. I have seen a most violent sympathetic fever, delirium, and death, follow, in consequence of
dividing parts in the leg, and searching after a bleeding artery. The loss of a limb above the knee is more than many can bear; the cutting for the stone, where it breaks, and
may be an hour in extracting, is also more than many can
bear; the parts being in such diseased state, as not to be relieved, have continued the symptoms of the disease; and
the loss of a testicle, although of so small a size, when compared with many other parts which we can lose with impu-
nity, yet from its vital connexion is more serious. We
cannot bear to lose much brain.

The loss of too much blood is often an attendant on, or
a consequence of operations; but sometimes takes place
without much violence. This produces very considerable
constitutional effects; bringing on weakness, and many
complaints, depending as it were, upon debility, which are
what are commonly called nervous. I have seen a locked
jaw come on in consequence of the loss of a considerable
quantity of blood, the cause of the loss being but trifling,
and giving no symptoms whatever.

The nature of the cause of inflammation, produces, I believe, but little variation in the constitution; for of whatever kind it is, the symptoms in the constitution will be in
all cases nearly the same, proportioned only to the violence
and rapidity of its progress; and as this inflammation is
pretty violent, more especially if it produces healthy suppuration, it generally produces more violent effects upon

Vol. II.   Q

the conftitution than any other; this however, will be in some degree according to the fufceptibility of the conftitution for inflammation ; and if any difference takes place in the inflammation in one conftitution from that of another, it will arife from the nature of the conftitution, the nature of parts, and their fituation, and not from the nature of the caufe.

The fympathy of the conftitution with a local difeafe, is what I have called univerfal fympathy, and is perhaps, the moft fimple act of a conftitution ; it is the fympathy, with a fimple violence, as a cold, etc. but ftill it will vary in different conftitutions, becaufe all conftitutions will not act alike under the influence of a local difeafe, although it will vary according to the ftages of inflammation, according to the natural difpofition of the parts inflamed, and the fitua-tion of thofe parts in the body ; yet it may be the moft fim-ple act of that conftitution at the time ; for although it would appear at the time to be an increafe of the difeafe by its becoming univerfal, yet as it is a natural confequence it is a much better fign of health, than if no fever had oc-curred in confequence of confiderable injuries ; for if there was no inflammation, there would probably be little or no fever.   Nature requires to feel the injury, for where, after a confiderable operation, there is rather a weak quiet pulfe, often with a nervous oppreffion, with a feeming difficulty of breathing and a loathing of food, the patient is in a dan-gerous way.   Fever fhews powers of refiftance, the other fymptoms fhew weaknefs finking under the injury.   This is like the effects of the cold-bath ; yet we fee it calling forth, or roufing up to action, fome peculiarity in the con-ftitution, or a part, which may be continued after the fympathetic action is loft, and which may again reflect back upon the part its reluctance to heal.   This may be examplified by affection or injury, fcrofula, even cancer, etc*.

* I believe that local fpecific irritations do not produce much variety in the conftitution; for I am perfuaded that fpecific local irritations are not capable of altering that conftitution, fimilar to the plague and other contageous difeafes.   I believe that mor-bid poifons do not act by any peculiar mode of action in the part, fo as to affect the conftitution in any peculiar way but fuch as are capable of continuing fo long as to weaken that con-ftitution, as for inftance, the lues, when of long ftanding;

Rigors are commonly the first symptom of a constituti-onal affection ; but a rigor is productive of other effects, or symptoms, as it were naturally rising out of the rigor ; and these are according to the nature of the constitution : in a strong constitution, a hot period succeeds, as if the consti-tution was roused to action to resist debility, which ter-minates the rigor ; and this hot fit terminates in perspira-tion, which is the complete action of the disease, re-storing tranquility, which is the cure, and is the best ter-mination that can happen where a rigor takes place ; for it shews that the constitution has the power of terminating the effects of the cause. I believe, however, that in most cases it shews a degree of weakness, especially if easily ex-cited, or a peculiarity of constitution. But as the cause is still continuing in cases of rigor arising from local irrita-tion, these rigors may recur ; and if they recur, it shews a constitution ready to be affected; however, if they do recur at stated periods, it still shews the constitution to be able to resist the effects of the disease. Further, if the constitution is weak, a rigor comes on, and no hot fit suc-ceeds, but it runs directly into the sweat ; it will proba-bly be cold and clammy. If it is a constitution of another kind, the hot fit will continue, having only a kind of abate-ment, but no sweat or perfect intermission will take place, and therefore the whole action has not taken place.

Rigors from local irritation, attended with the full action, and at regular stated times, have all the characters of an intermittent fever ; but it may be observed that, in common, rigors preceding suppuration are not followed by so much heat and sweating, as an intermittent is.

In spontaneous inflammations, it is not so easy to ascer-tain, whether the constitution or the part is first affected, and if it always could, it would be the best guide to know whether the inflammation was local entirely, or an effect of a constitutional affection ; nothing but the priority of

but this will be similar to every other lasting disease : for at first it certainly does not affect the constitution so as to alter the dis-position of a wound made upon any part. I am not so certain respecting natural poisons. The tecuna, poisoned arrow, etc. would seem to produce a peculiar constitutional affection, from a local cause ; for we can hardly suppose absorption to have ta-ken place in so short a time.

the fymptoms can in fome degree fix this ; but the conftitutional fymptoms are often of flight, at leaft at firft, as not
to be taken notice fo.   However, we know that indifpofitions of the conftitution are productive of local complaints,
which are often attended with inflammation, but which is
often according to the nature of the parts*, the conftitution
being firft difeafed ; and we know that in many fevers
there is fuppuration in fome part of the body, and often in
particular parts, fuch as the parotid glands, probably according to the nature of the fever ; fuch inflammations will,
according to their violence, add to the conftitutional affection.   Conftitutional affection arifing from inflammations
will be almoft coeval with the inflammations, or at leaft
will very foon follow; however, that will be according to the
circumftance before related ; for inflammation is an act of
the part, attended with a degree of violence, and the conftitution will feel it fooner or later, according to circumftances : we fee in cafes of inflammation of the tefticles from
a gonorrhœa, (which muft be confidered as entirely local)
that the conftitution is foon affected by it.   But conftitutional fymptoms arife from external violence alone, and
more efpecially when attended with lofs of fubftance ; and
they will be fooner or later, according to the degree of the
violence, and the importance of the part loft agreeably to
what has been faid ; but fimple violence, even with the
lofs of a part, I have already obferved, is not of fuch confequence as we fhould at firft imagine : for in confequence
of the lofs of a limb, if the parts are allowed to heal by
the firft intention, the conftitution is but little affected ;
it is, therefore, violence with lofs of fubftance, and which
is to produce inflammation and fuppuration, that gives rife
to the conftitutional fymptoms; and when thefe commence,
or, more probably, when the part fets about thefe operations, the conftitution becomes affected.   It is more the
new difpofition in thefe parts, than the quantity of inflammatory action in them, by which the conftitution is affect-

---

* Local inflammations arifing from derangement of the conftitution, I think are moft commonly of the fcrofulous kind,
more efpecially when in parts of a particular nature, fuch as
lymphatic glands, ligamentous or tendinous parts. which, when
in particular fituations, are often fuppofed to be venereal.   Vide Treatife on Venereal Difeafe.

ed ; for we shall see, that upon the simple commencement
of the suppurative disposition, before it has taken place,
rigors, etc. come on.

The constitutional effects arising upon the commencement of inflammation independent of situation, of vital
parts, nerves, etc. are greater or less according to the nature of the disease. When the adhesive stage commences,
it has but very little effect upon the whole system ; there
is sometimes however, a rigor, although not always ; this
is more in common spontaneous inflammations than in those
arising from an injury done to a part, but such are seldom
or ever alarming. When the suppurative disposition takes
place, new effects upon the constitution arise, which are
very considerable and varying in themselves. The cold fits,
or rigors, are more frequently felt at the commencement of
the suppurative than at the beginning of the adhesive inflammation, more especially too if it is what we commonly
call spontaneous inflammations, which advance to suppuration ; for in those inflammations occasioned by an accident,
or an operation, which must suppurate, they appear to set
out at the very first with a kind of suppurative disposition.
Those arising in consequence of spontaneous inflammation,
or from an injury, are not lasting, are often succeeded by
hot fits, and if they terminate in perspiration then the patient is relieved ; and are more or less so according to the
greatness of the present inflammation and the suppuration
that is likely to follow, joined to the nature of the parts and
their situation : if in vital parts they will be most violent,
and next to these, in parts far from the heart. This cold
fit is, indeed, a constant symptom in most local diseases,
which affect the constitution ; and in this case, plainly
shews that the constitution is so affected, or sympathizes
with the part. It is thus also, that fevers usually commence,
and upon the absorption of any poisonous matter the same
symptoms appear. I have seen them arise from a simple
prick in the end of the finger, made with a clean sewing
needle*, exactly similar to those arising from the absorption of
poison. Disagreeable applications to the stomach produce
them, and also disagreeable affections of the mind: but rigors are not confined to the commencement of disease, for

* Hence it would seem as if simple irritation in a part, was
capable of affecting the whole nervous system.

they occur in its progrefs, and fometimes at its termintaion, as will be mentioned.

It is probable that the ftomach is the caufe of thofe rigors, by its taking part in the difeafed action of the conftitution ; for as the ftomach is the feat of fimple animal life, and thereby the organ of univerfal fympathy of the materia vitæ, or the living principle, it is of courfe more or lefs affected upon all thefe occafions ; fo that an affection of any part of the body and of the mind, can produce very nearly the fame effect, as that which arifes from difagreeable applications to the ftomach itfelf ; which accounts for that vifcus taking part in all conftitutional affections. I am inclined to believe, that fympathy of the ftomach which occafions ficknefs, arifes from caufes producing weaknefs or debility.   It takes place from injuries or diforders of the brain, which occafions univerfal debility ; it arifes from lofs of blood, and alfo from epileptic fits.   How far the ficknefs is to be confidered as an effect which is to produce action, viz. vomiting and which action is to reflect ftrength back upon the conftitution, I do not know ; but it is certain, that people who are fick, and going to faint, are prevented by the action of vomiting ; the act of vomiting, therefore, appears often to be a caufe of the prevention of the fits coming on, by roufing up the actions of life.   The rigors I fhould be apt to fufpect arife from weaknefs at the time.   A fudden alteration, a fudden call, or a fudden and univerfal irritation upon the conftitution, will, I imagine, produce immediate weaknefs ; for every new action in a conftitution, muft produce or tend to produce a weaknefs in that function ; the effects of which will vary according to the neceffity, and ftate of the conftitution. In fome cafes where the conftitution is ftrong, and as it were equal of itfelf to the tafk, it will call up the animal powers to action, and produce the hot fit of a fever ; but in weak conftitutions, or in fuch as threaten diffolution, as in many difeafes, efpecially towards the clofe, it lofes by every rigor, and is feldom capable of producing a hot fit, but only occafions a cold clammy fweat ; hence, cold fweats, when a perfon is in extremities, is a common fymptom.   That rigors are an effect of every fudden change in the conftitution, and are not peculiar to the commencement of difeafe, is evident from the following cafes ; which alfo prove, that even the change to health fhall produce the fame effect, fo that not only in its commencement, and in its different stages, a disease shall produce rigors, but in its termination or crisis.

A boy about eleven months old was taken ill with a complaint, which could not be well understood from the symptoms, and which came on insensibly. His pulse was quick and full, for which he was bled three times, and the blood was rather sizy; the tongue was white; he was not very hot, but uneasy and restless, with loss of appetite. His stools were upon the whole pretty natural; he was observed to be every other day rather worse, although there never was a perfect intermission, but only a kind of remission. After having been ill for about a fortnight in this way, he was taken with a cold shivering fit, succeeded by a hot fit, and then a sweat. My opinion was, that the disease was now formed, and that he would have more at the intermitting times; but he had no more after. In short, the disease formed itself into that which has but one fit, and in this formation he had those symptoms. I have seen the same symptoms in many diseases, especially those occasioned by an operation, which in general alarm, but which should not if they go through their stages. A patient of mine at St. George's Hospital was cut for the stone; he had no uncommon symptoms for several weeks, when he was taken with a cold fit, which was succeeded by a hot one, and then by a profuse sweat. The young gentlemen of the hospital were rather alarmed, conceiving them to be the signs of dissolution; but I told them that this was of no consequence, as the disease had completed its full action. That it was either a regular ague, or arose from the irritation of the wound; and if the first, he would have more of them at stated periods, which the bark would probably cure; but if the second, it might not return; for, since the constitution was in possession of the complete action, that when the parts got better he would be well. He had no more; and went on doing as well as if no such fit had ever taken place. This is not the only instance of this nature.

Here it is to be considered, that those affections of the constitution, are effects of the local action of the solids, either when produced by spontaneous causes, or by accident; but there are sometimes constitutional symptoms, or universal sympathies, which arise immediately out of the act of the violence itself, and which are often dangerous. Loss of blood may be reckoned one cause, which will bring on all kinds of conflitutional complaints, in confequence of weaknefs being produced, either immediate, as fainting, or fecondary, as in dropfies, as well as nervous affections ; the locked jaw, for inftance ; or violence alone without the lofs of blood, may often produce immediate fatal effects.

I have feen a man thrown into fuch convulfions from the operation of the hydrocele being performed upon him, that I began to defpair of his recovery. I have known a man die immediately of caftration. Thefe fymptoms are fomewhat fimilar to the fecond, or nervous, but are ftill very different ; for in the prefent, the perfons are as it were loft to themfelves, being rendered fenfelefs, therefore, it is probably more an affection of the brain than the nerves.

Another fymptom attending inflammation when it has affected the conftitution, is frequent exacerbations, or periods in which the inflammation appears to be increafed. They have great affinity to the rigors we have been mentioning.

Exacerbations are common to all conftitutional difeafes, and would often appear to belong to many local complaints. They are commonly regular if the conftitution is ftrong, having their ftated times, and in proportion as they are fo the difeafe is lefs dangerous. They are a repetition of the firft attack, but feldom fo ftrong, except where there is a perfect ceffation in the difeafe between the fits. This is an attribute belonging to life; and fhews that life cannot go on the fame continually in any ftate, but muft have its hours of reft, and hours of action.

In this, as in almoft every other fymptom of difeafe, the effect has been confidered as a caufe ; for exacerbations have always been confidered as owing to the difeafe having its time of fubfiding, or leffening, and its time of increafe. This idea might pafs as juft in fevers where caufes are not known ; but where the caufes continue the fame, as in local difeafes, we, à priori, fhould not expect it ; yet we find in fuch cafes, periods of increafe, and decreafe of the fymptoms, in the conftitution, and therefore we muft fearch after fome principle belonging to animal life, as a caufe of this.

We fhall find that an animal is fo conftituted as to be incapable of exifting for any continuance of time in any one state whatever ; the actions of the senfitive principle, when in perfect health, have their regular exacerbations, viz. watching, and sleep ; it is difeafe that interrupts this regularity of the actions of health ; therefore we find that the actions of difeafe cannot always go on in the fame way ; nature refts infenfible of the difeafe, while the difeafe exifts at all times alike : fince this is the cafe, where we fee evidently a continuance of the remote caufe, and that the conftitution is only capable of being affected by this caufe at ftated times ; according to the fpecies of irritation given, and the conftitution at the time ; may we not reafonably fuppofe this to be the cafe, where the caufe is invifible, as in fevers.

Whether thefe exacerbations are an effect of an occafional increafe of the inflammation, or whether the inflammation is increafed by the paroxyfm of the fever, is not eafy to determine ; but they attend each other.

An ague is a difeafe which exifts in the conftitution, between the fits, as much as at the time of the fit ; but the conftitution becomes infenfible of it, and the action can only laft a ftated time.

The procefs of ulceration feldom appears to affect the whole fyftem ; it is hardly known to exift, but in the appearance of the parts, viz. when the part which contains the matter gets fofter to the touch, or when an ulcer becomes larger. But that rigors take place upon the commencement of ulceration, I think is evident ; although it cannot well be known in all cafes ; for ulceration will be fo clofe upon fuppuration in moft cafes, that it will be difficult to diftinguifh which was the caufe of the rigor ; but where fuppuration has taken place, and the abfcefs is opened, fo that the firft act of fuppuration is finifhed, yet if it is not opened fo as to allow of a ready outlet to the matter ; for inftance, if not opened at a depending part, the preffure of the matter againft the moft depending part of the abfcefs will produce ulceration there, and rigors will take place. Thofe rigors, however, will not commence for fome time after the firft opening, becaufe the firft opening will for fome time remove the difpofition for ulceration all over the furface of the abfcefs ; but when it finds that this opening is not fufficient to take off the preffure, then it fets

Vol. II.   R

about forming another opening, and when it does so, the rigors will recur, and with as much severity as before. This is supposed by some to be new matter forming from fresh inflammation, and by others to be the absorption of matter already formed. Although ulceration does not affect the constitution equal to the mischief it is doing, yet its operations are often much affected by indispositions of the constitution ; in some indispositions its progress is increased, in others it is even brought on, as in many old sores, especially of the lower extremities ; and in some indispositions its progress is leffened or stopped.

The constitutional symptoms arising from a local complaint may be divided into three as to time, the immediate, indefinite, and remote. Of the first, or immediate, there appears to be but one ; of the second, there is probably a great variety, at least, appearing in very different forms and at very different periods, in respect of the original cause. Of the remote there is probably only one. The immediate I shall reckon that which is called, the symptomatic fever ; and what I shall reckon the second are, nervous affections, as spasms, both temporary and permanent, and delirium. Whether the symptomatic fever, the spasms, or the delirium, come first, is not certain, for often all concur or occur at the same time ; but as the sympathetic fever is most constant, and is more an universal principle, it is to be reckoned the first. And the third, which I have called, remote, is what is understood by the hectic ; to which may be added the symptoms of dissolution, which is the last stage of all, and may be a consequence of either the above, or any other disease.

The first of the constitutional affections is commonly called, the symptomatic fever ; but which I choose to call the sympathetic inflammatory fever. This is immediate, or nearly so, and is the sympathy of the constitution, with the first stages of a local disease, which excites an alarm in the constitution, thereby rousing up its powers to produce succeeding actions. This would appear to show very much the nature of the constitution at the time ; for not being of any specific nature, both inflammation and fever are led of course into the nature of the constitution by the natural tendency of the constitution itself, and therefore partake of it, and only become more or less of a specific, in proportion as the conftitution has more or lefs of a fpecific fufceptibility or difpofition.

I have already obferved, that affections of the conftitution often commence with rigor. However, the commencement of the fympathetic fever is not always attended with that effect; and I believe it is the beft conftitution where it is not; and in that cafe, it changes into a regular fever of the inflammatory kind. If the conftitution has powers, heat comes on, attended with dry fkin, frequent and commonly a full pulfe, having at the fame time a degree of hardnefs in the ftroke; watchfulnefs, high coloured urine, lofs of appetite for folids, and thirft; all thefe will vary according to various vifible circumftances, as well as according to many invifible ones, fome one fymptom being more in one conftitution, and lefs in another.

It is in many inftances difficult to determine what is caufe, and what is effect. It has been commonly fuppofed that this fever was necefiary for the operation of fuppuration, and therefore the fever did not arife from the fympathy of the conftitution with a local injury, but as a neceffary effect to become a caufe of fuppuration. If this was the cafe, we could have no fuppuration which had not been preceded by fever; and the fever muft have been equal in all cafes in the fame conftitution, let the quantity of injury be what it will. For if a pimple, or the fuppuration of a fcratch depended upon fever, they would require as much fever for their production of inflammation and fuppuration as the largeft abfcefs, or largeft wound; for a point that inflames and fuppurates is under the fame predicament with refpect to the whole that a thoufand are; and a large abfcefs is to be confidered as only made up of a thoufand points. One venereal fore requires as much mercury to cure it as a thoufand. One plant requires as much wet weather and funfhine as a million. A principle that affects univerfally can only affect a part in proportion to the quantity of the univerfal affection there is in the part, each part has juft its portion of general influence.

Now, according this propofition, which is undeniable, a fcratch requires the fame quantity of fever that an amputation of the thigh does. Let us fee how this accords with common experience; we find that inflammations and fuppurations of fores fhall take place without any fever ; that the fever, in confequence of an injury, is not in all cafes in the leaft proportioned to the quantity of injury, inflammation, and fuppuration, which it always fhould, if the laft was an effect of it ; and we know if an increafe of fever comes on, fuperadded to the fympathetic, that fuppuration is retarded or ftopped altogether, inftead of being quickened.

From the fame mode of reafoning it fhould be exactly the fame whether the fever produce fuppuration in a vital part or not a vital part. It is much more eafy to conceive that an injury done to vital part fhall be the caufe of univerfal fympathy, than that a vital part fhould require more fever to make it inflame and fuppurate than a part does which is not vital. This theory would at once overfet our obfervation that the conftitution is affected or fympathizes more readily with fome parts than with others. In many cafes of fpontaneous inflammations and fuppurations it was natural to fuppofe that the fever was the caufe of the fuppuration ; but if perfons who thought fo had obferved accurately, they would have divided fpontaneous fuppuration into two kinds ; one, whofe remote and immediate caufe was local, and therefore in fuch the fever followed the local action, as in injuries ; the other, where the remote caufe was fever, which produced the injury ; and the injury, whatever it was, produced the inflammation and fuppuration ; fo that here fever preceded, and was neceffary for the remote caufe, but not as the immediate ; and indeed, as a proof of this, fuppuration hardly takes place till the fever is gone. The fmall-pox is of this laft kind ; as probably many other contagious difeafes. ·

Thofe fymptoms continue more or lefs, according to the degree of injury, the nature and fituation of parts, and the conftitution ; but as they arife from a local caufe, which fubfides, they of courfe fubfide alfo ; however, as the conftitution has often an inflammatory tendency, or a tendency to fome other difeafe, befides the action arifing from the violence fingly, the parts often run into it, and this is reflected upon the conftitution, which paffes into that action to which it has a tendency, by which fever is kept up, and thereby inflammation.

The fubfiding of thefe fymptoms is the cure ; and where they are fimply the effects cf the violence, the fever cures itfelf ; therefore, the only thing neceffary is to leffen its violence ; but if the injury is of any fpecific kind, that fpecific quality muft be corrected, if poffible, and then the cure will take place.

As the motion of the blood in the whole fyftem is in- creafed, and as we have reafon to fuppofe it is locally in- creafed, then what will diminifh the motion of the blood, will relieve in this refpect; there are two methods of doing this ; the firft, by taking off its force ; and this will be ef- fected by bleeding.

This, if it does not leffen its motion, or take off from the fympathy of the conftitution, with the local difeafe, yet it leffens the momentum in the whole and in the part, which is taking off the effect of the excefs of motion in the blood.

The other is diminifhing the action of the parts by affect- ing the conftitution, which may be done by purging ; in this light, bleeding may alfo be in fome degree confidered. It becomes in fuch cafes very neceffary to relieve the con- ftitution by leffening the action of that conftitution ; for although what has been advifed, was to leffen the inflam- mation itfelf, and thereby leffen its effects on the conftitu- tion, yet as that feldom is done fufficiently to remove any affection of the conftitution, we muft therefore pay atten- tion to that conftitution ; the two remedies will in fome degree go hand in hand, one affifting the other ; for in- ftance, in a ftrong healthy conftitution where the fympto- matic fever runs high, bleeding and purging will have their double effects , but ftill the conftitution may require its peculiar medicines, which will in a fecondary way re- lieve the inflammation.

The fecondary conftitutional fymptoms are not fo deter- mined as to time ; I have called them nervous, although not ftrictly fo in every cafe, becaufe more variety of affec- tions are produced than from any caufe I know ; yet thefe affections feem all to have more connexion with the nervous than the vafcular fyftem, and are feverally excited by the particular tendency or fufceptibility of different conftitu- tions. Many of them, I believe, are more common to the young than the old, which come under the doctrine of univerfal nervous fympathy, with a local complaint ; of this kind are univerfal convulfions from teething or worms; local convulfions, as St. Vitus's dance, and probably ma-

ny others not so well marked, as those which worms and
teething often produce. I have seen hickup come on early
in consequence of an operation; but in this stage of the nervous affection, little was to be apprehended, although it
certainly shewed a peculiarility of constitution, and such as
should be attended to ; but when hickup occurs towards
the last stages, it shews strong signs of dissolution.

' Many full-grown persons are also subject to very severe affections of the nervous kind, especially those people who are
called nervous ; and more particularly still those who have
bad affections, in consequence of complaints of the stomach.
In such constitutions there is observed great dejection,
sinking, cold sweats, hardly any pulse, loss of appetite, no
sleep, etc. seeming to threaten dissolution ; those symptoms
are worse by fits. Delirium appears to arise from nervous
affection of the brain, or sensorium, producing a sympathy
of the action of the brain, with the materia vitæ of the
parts ; not sensation as a head-ach, but action, producing
ideas without the exciting impression, and therefore delu-
sive. This symptom is common to them all ; it is frequently a consequence of their being violent, or carried to
considerable length in their several kinds : often arising in
consequence of compound fractures, amputation of the
lower extremities, injuries done to joints, brain, etc. but
not so often attending the hectic, although it is often a
symptom of dissolution. We have agues also from many
diseases of parts, more especially of the liver, as also of the
spleen, and from induration of the misenteric glands.

The following cases are remarkable instances of well
marked constitutional diseases from local irritation, where
the constitution took on a particular action, to which it had
a strong tendency. A gentleman had a very bad fistula in
perinæo from a stricture, and when the water did not come
freely, an inflammation in the part and scrotum was produced, and then he had an ague, which was relieved for a
time by the bark. Two children had an ague from worms,
which was not in the least relieved by the bark, but by destroying the worms they were cured.

As these diseases which I have brought into this class
are of such various kinds, each must be taken up apart, and
treated accordingly ; but they are such as yield very little to
medicine, for in some the constitutional disease is formed,
and does not require the presence of the local disease to keep

it up, as in the tetanus; and in others, the local difeafe
being ftill in force, it is not to be expected, that the conftitutional affection is to be entirely relieved, although in fome
degree it may. In thofe which form a regular conftitutional difeafe, fuch as an ague, although the local difeafes
may ftill exift in full force, yet fome relief may be expected; the bark is to be adminiftred, although not with a view
to cure, as the immediate caufe ftill exifts; but bark will in
fome leffen that fufceptibility in the conftitution, and may
cure at leaft for a time, as I have feen in agues arifing from
the fiftula in perinæo. But the fufceptibility in the two
children, cited above, was fo ftrong for fuch a difeafe, that
the bark was not fufficient; and therefore, when the local
caufe is not known, and when the common remedies for
fuch effects do not cure them, fome local difeafe fhould be
fufpected. We fee often fuch fymptoms arifing from difeafes of the liver, and the bark curing this fymptom, yet the
liver fhall go on with its difeafe, and probably fafter, as I
believe bark is not a proper medicine for difeafes of this
vifcus; fuch complaints of the liver have been too often attributed to the curing of the ague improperly by bark. St.
Vitus's dance, and many other involuntary actions, have a-
rifen from the fame caufe; fuch conftitutions required only an immediate caufe to produce the effects. It it poffible, however, that no other mode of local irritation would
have produced the fame effect, every conftitution having a
part that is capable of affecting it moft. We find alfo local
effects in confequence of local injuries, as the locked jaw,
etc. which are remote fympathies with the part affected,
which may become pretty univerfal, and which cannot be
called immediate effects as to time, as they are often forming after the fympathetic fever has taken place, efpecially
the locked jaw, which appears in many cafes to be formed
in the time of the preceding difeafe, and not appearing till
it had fubfided. There are certain intermediate fteps between the inflammatory and the hectic ftate; but neither
cure nor diffolution take place in this period.

The following cafe illuftrates the effects of inflammation
on the conftitution.

A lady, of what is called a nervous conftitution, arifing in
fome degree from an irritable ftomach, often troubled with
flatulencies, and what are called nervous head-achs, with

pale urine at thofe times, uncomfortable feelings and often
finkings, had a tumor removed from the breaft, and likewife from near the arm-pit ; nothing appeared uncommon
for a few days, when very confiderable diforders came on.
She was attacked with a fhivering or cold fit, attended with
the feel of dying, and followed with cold fweat. It being
fuppofed that fhe was dying, brandy was thrown in, which
foon brought on a warmth, and fhe was relieved ; the fits
came on frequently for feveral days, which were always
relieved by brandy ; and fhe took in one of the moft violent of them about half a pint of brandy.

While under thefe affections fhe took the bark as a
ftrengthener ; the mufk, occafionally, as a fedative in pretty large quantities ; camphorated julap frequently, as an
antifpafmodic ; and towards the laft fhe took the valerian
in large quantities : but whatever cifect thefe might have
in leffening the difeafe on the whole, they were certainly
not equal to it without the brandy. Brandy removed thofe
dying fits, and I thought they became lefs violent after taking the valerian.

A queftion naturally occurs ; would the brandy alone, if
it had been continued as a medicine, have cured her, without the aid of the other medicines ? The other medicines,
I think, certainly could not have done it ; nor do I believe
that the brandy could have been continued in fuch quantity as to have prevented their returns, if fo, then the two
modes were happily united, the one gradually to prevent,
the other to remove immediately the fits when they came
on.   This cafe from the general tenor of the conftitution,
was running with great facility into the hectic.

CHAPTER V.

OF PUS.

Hitherto I have been treating of the operations of parts, preparatory to the formation of pus ; I am now come to the formation of that fluid, its nature and fuppofed ufes.

The immediate effect of the mode of action above de-fcribed, is the formation of a fluid, commonly termed pus ; this is very different from what was difcharged in the time of the adhefive ftage of the inflammation, when either form-ed in the cellular membrane or circumfcribed cavities ; it is alfo very different from the natural fecretion of internal canals, though it is probably formed in both by the fame veffels, but under very different modes of action.

The cellular membrane, or circumfcribed cavities, have their veffels but little changed from the adhefive ftate at the commencement of the fuppurative difpofition ; fo that they ftill retain much of the form they had acquired by the firft ftate, the difcharge being at the beginning little more than coagulating lymph mixed with fome ferum. This is fcarcely different from the adhefive ftage of the inflamma-tion ; but as the inflammatory difpofition fubfides, the new difpofition is every inftant of time altering thofe veffels to their fuppurative ftate ; the difcharge is alfo varying and changing from a fpecies of extravafation to a new formed matter peculiar to fuppuration; this matter is a remove fur-ther from the nature of the blood, and becomes more and more of the nature of the pus ; it becomes whiter and

whiter, lofing more and more of the yellow and green,
which it is apt to give the linen that is ftained with it
in its firft ftages, and in confiftence more and more vifcid,
or creamy.

By the formation of this new fubftance, the coagulating
lymph, which was extravafated in the adhefive ftate of the
inflammation, and adhered to the fides of the cells, either in
cut furfaces as in wounds, in abfceffes, or circumfcribed cavities, is pufhed off from thefe furfaces, and if it is the
inner furface of a cavity, it is pufhed into it, fo that the
cavity contains both coagulating lymph and pus ; or if it is
a cut furface, the coagulating lymph is feparated from it
by the fuppuration taking place, and is thrown off; but as
fuch furfaces are generally dreffed immediately after the
operation, while the wound is bleeding, this blood unites
the dreffings to the fore, which is affifted afterwards by the
coagulating lymph thrown out in the adhefive ftage, the
whole, viz. dreffings, blood, and coagulating lymph are
generally thrown off together, when fuppuration commences on thefe furfaces. This is the procefs which takes
place in the firft formation of an abfcefs, and the firft procefs towards fuppuration in a frefh wound.

Upon the internal furfaces of the canals, the parts do
not go through all thofe fteps ; they would appear to run
into fuppuration almoft inftantaneoufly ; however, inflammations even here is a kind of forerunner of fuppuration·
This difcharge, from internal canals, has never been reckoned true matter, it has been called mucus, etc. but it
has all the charactere of true pus, which I am yet acquainted with.

Pus is not to be found in the blood, fimilar to that which
was produced in the firft ftage ; but is formed from fome
change, decompofition, or feparation of the blood, which
it undergoes in its paffage out of the veffels, and for effect-
ing which the veffels of the parts have been formed, which
produces a fubfiding of the inflammation from which it took
its difpofition ; hence it muft appear, that the formation
of pus confifts of fomething more than a ftraining of juices
from the blood. Many fubftances indeed which are to be
confidered as extraneous bodies in the blood, being only
mixed with, and not making an effential part of that fluid,
and perhaps even neceffary to it, may pafs off with the pus,
as with every other fecretion, yet the pus is not be confi-

dered on that account as simply parts of the blood unchang-
ed; but we must look upon it as a new combination of
the blood itself, and must be convinced that in order to
carry on the decompositions and combinations necessary
for producing this effect, either a new or peculiar structure
of vessels must be formed, or a new disposition, and of
course a new mode of action of the old must take place.
This new structure, or disposition of vessels, I shall call
glandular, and the effect or pus, a secretion.

I OF THE GENERAL OPINION OF THE FORMATION OF PUS.

THE dissolution of the living solids of an animal body
into pus, and that the pus already formed has the power
of continuing the dissolution, is an old opinion, and is still
the opinion of many; for their language is, " Pus corrodes,
" it is acrid, etc." If this idea of theirs was just, no fore
which discharges matter could be exempted from a conti-
nual dissolution; and I think it must appear inconsistent,
that the matter which was probably intended for salutary
purposes, should be a means of destroying the very parts
which produced it, and which it is meant to heal. Probably they took their idea from finding, that an abscess was
a hollow cavity in the solids, and supposing the whole of
the original substance of this cavity was now the matter which
was found in it. This was a very natural way of accounting for the formation of pus, by one entirely ignorant of
the moving juices, the powers of the arteries, and the operation of an abscess after it was opened; for the knowledge of these three, abstracted from the knowledge of the
abscess before opening, should have naturally led them to
account for the formation of pus from the blood by the
powers of the arteries alone; for upon their principle these
abscesses should continue to increase after opening, as fast
as before. Upon this principle being established in their
minds; viz. that solids were dissolved down into pus,
they built a practice which was to bring all indurated parts
to suppuration if possible, and not to open the suppuration

in such parts early ; this was done with a view to give such
solids time to melt down into pus, which was the expref-
fion; but according to their own theory, they seemed to for-
get that abfceffes formed matter after opening, and therefore the parts ftood the fame chance of diffolution into pus
as before. Alfo, from being poffeffed of this idea, that
folids went into the compofition of pus, they never faw pus
flowing from any internal canal, as in a gonorrhœa, etc.
but they concluded that there was an ulcer ; we would for-
give fuch opinions, before the knowledge that fuch furfaces could and generally did form pus without a breach
of the folids ; but that fuch an opinion fhould exift afterwards is not mere ignorance, but ftupidity ; and the very
circumftance of internal circumfcribed cavities, as the abdomen, thorax, etc. forming pus, where they might often have
feen pints of matter, and yet no breach in the folids to have
produced it, which is a proof beyond controverfy, fhould
have taught them better ; fuch ideas difcover defect of
knowledge and incapacity for obfervation.

The moderns have been ftill more ridiculous, for knowing that it was denied, that folids were ever diffolved into
pus, and alfo knowing that there was not a fingle proof of
it, they have been bufy in producing what to them feemed proof. They have been putting dead animal matter into abfceffes, and finding that it was either wholly or in
part diffolved, they therefore attributed the lofs to its being formed into pus ; but this was putting living and dead
animal matter upon the fame footing, which is a contradiction in itfelf ; for if the refult of this experiment was really according to their idea of it, the idea of living parts
being diffolved into pus muft fall to the ground, becaufe
living animal matter and dead animal matter can never
ftand upon the fame ground.

Common obfervation in their profeffion fhould have
taught them, that even extraneous animal matter would lie
in abfceffes for a confiderable time before it was even diffolved. They might have obferved in abfceffes arifing either from violence, or from a fpecies of eryfipelatous inflammation, that there were often floughs of the cellular
membrane, and that thofe floughs would come away like
wet tow, and therefore were not diffolved into pus,

They might alfo have obferved in abfceffes on tendinous
parts, as about the ancle, etc. that often a tendon became

dead and floughed away, and that thefe fores do not heal
till fuch parts have floughed, and this is often not accom-
plifhed for months, and yet all this time thofe floughs are
not formed into pus. They might have alfo known, or
obferved, that pieces of dead bone fhall lie foaking in matter for many months, and yet not diffolve into pus ; and although bones in fuch fituations fhall lofe confiderably of
their fubftance (which might by the ignorant be fuppofed
to have been diffolved into pus) yet that wafte can be accounted for and proved on the principle of abforption ; for
they always lofe on that furface where the continuity is broke off, and which is only a continuation of the feparating procefs.* To fee how far the idea was juft, that
dead animal matter was diffolved by pus, I put it to the
trial of experiment, becaufe I could put a piece of dead a-
nimal matter of a given weight into an abfcefs, and which
could at ftated times be weighed ; to make it ftill more fa-
tisfactory, a fimilar piece was put into water, kept to nearly the fame heat : they both loft in weight, but that in the
abfcefs moft, and there was alfo a difference in the manner,
for that in the water became fooneft putrid ; but thefe experiments having been made as far back as the year 1757,
I fhall not rely on their accuracy, but ftate them as made
by my brother-in-law, Mr. Home, and, as given in his
Differtation on the Properties of Pus, page 32, under the
idea that pus had a corroding quality.

" As pus has been fuppofed to have a corroding quality,"
I may add even upon the living folids, " I made the following experiments to afcertain the truth, or falacy of fuch
an affertion, and found it to be void of foundation, and to
have arifen from the inaccuracy of obfervers having prevented them from feeing the diftinctions between pus in a pure
ftate, and when mixed with other fubftances.

" EXPERIMENT.

" I made a comparative trial upon matter contained in
an abfcefs, and on pus and animal jelly out of the body.
The matter and jelly were in equal quantities and contai-

* It may be fuppofed that bones are not capable of being diffolved into pus ; but we know that bone has animal fubftance
in it, and we alfo know that this animal fubftance is capable of
being diffolved into chyle.

ned in glafs-veffels, kept nearly in the temperature of the
human body. To make the comparative trials as fair as
poffible, a portion of mufcle, weighing exactly one drachm,
wasimmerfed in the matter of a compound fracture in the
arm of a living man, and a fimilar portion into fome of the
fame matter out of the body; alfo a third portion into fluid
calf's foot jelly, in which the animal fubftance was pure,
having neither wine nor vegetables mixed with it. Thefe
three portions of mufcle were taken out once every twenty-
four hours, wafhed in water, weighed and returned again.
The refults were as follows:

" In twenty-four hours.—The portion of mufcle in the
abfcefs weighed fixty grains, was pulpy and foft, but quite
free from putrefaction: that portion immerfed in the pus,
weighed forty-fix grains, was pulpy, foft, and had a flightly putrid fmell: the portion in the jelly weighed thirty-
eight grains, was fmaller and firmer in its texture.

" Forty-eight hours.—The portion of mufcle in the abfcefs weighed thirty-eight grains, and had undergone no
change: that in the matter weighed thirty-fix grains, was
fofter and more putrid: that in the jelly thirty-fix grains
and fmaller.

" Seventy-two hours.—The portion of mufcle in the abfcefs weighed twenty-feven grains, was drier and firmer:
that in the matter eighteen grains, and was rendered fibrous
and thready: that in the jelly unaltered.

" Ninety-fix hours.—The portion of mufcle in the abfcefs
weighed twenty-five grains: that in the matter was diffolved: that in the jelly weighed thirty-fix grains*.

" One hundred and twenty hours.—The portion of mufcle in the abfcefs weighed twenty-two grains, not at all putrid: that in the jelly thirty-four grains, not at all putrid.

" One hundred and forty-four hours.—The portion of
mufcle in the abfcefs weighed twenty-two grains, and was

---

* One reafon probably for the piece of meat fo foon becoming putrid and diffolving in the pus, was its being kept in the
fame pus the whole time; therefore its diffolution was owing
more to putrefaction than a diffolving quality in the pus; whereas the piece in the abfcefs had its matter continually changing,
which is the common refult in a fore, and if it had a corroding
quality independant of the putrefaction, it ought to have b en
diffolved firft; but we may obferve that, the piece of mufcle in
the abfcefs, and the piece in the jelly were nearly upon a par.

free from putrefaction: that in the jelly thirty-four grains."

The fuppofed facts of the folids diffolving being eftablifhed in the mind as fo many data to reafon from, they had now no difficulty to account for the formation of pus from both the folids and the fluids; fermentation ftarted up in the mind immediatly as a caufe; but there muft be a caufe for fermentation; and according to this idea, there are facts which go againft it: firft, let us confider internal canals, where only mucus is naturally formed, taking on the formation of pus without the lofs of fubftance, or any previous ferment, and leaving it off.

Now if a fermentation of the folids and fluids was the immediate caufe, I fhould beg leave to afk what folids were deftroyed in order to enter into the compofition of the pus difcharged; for the whole penis could not afford matter enough to form the pus, which is difcharged in a common gonorrhœa; I fhould alfo beg leave to be informed, how that fermentation of the fluids ever ceafed, for there is the fame furface, fecreting its mufcus, whenever the formation of pus ceafes.

Befides, if diffolved folids enter neceffarily into the compofition of pus, by the power of fome ferment, it may be afked by what power the firft particle of this fluid in an abfcefs or fore is formed, before there is any particle exifting which is capable of diffolving the folids?

An abfcefs fhall form, and fuppuration ceafing, it fhall become ftationary, perhaps for months, and at laft be abforbed, and the whole fhall heal; what becomes of the ferment the whole time it is ftationary?

It has been fuppofed that blood when extravafated becomes of itfelf pus; but we find blood, when extravafated, either from violence or a rupture of a veffel, as in an aneurifm, never of itfelf becomes pus; nor was pus ever formed in fuch cavities till inflammation had taken place in them, and that in fuch cavities there was to be found both the blood and the matter; if the blood had coagulated, (as it feldom does in thofe cafes of violence) it would be found ftill coagulated, and if it had not coagulated the pus would be bloody.

True pus has certain properties, which when taken fingly may belong to other fecretions, but when all joined, form the peculiar character of pus, viz. globules fwimming in a fluid, which is coagulable by a folution of fal ammoniac,

which no other animal fecretion that I know is; and at the
fame time a confequence of inflammation; thefe circumftances taken together may be faid to conftitute pus.

As inflammation does not produce at firft true pus, I
made the following experiments to afcertain its progrefs or
formation. To do this it was only neceffary to keep up an
irritation on fome living part a fufficient time to oblige it
to fet about the natural confequent actions, and the fmooth
coat of an internal cavity appeared to me to be well calcula-
ted for fuch an experiment, where nothing could inter-
fere with the actions of the parts, or their refult, and it
would alfo fhew its progrefs on internal furfaces, which
fhows its progrefs in wounds and abfceffes.

II. EXPERIMENTS TO ASCERTAIN THE PRO-GRESS OF SUPPURATION.

EXPERIMENT I.

THE tunica vaginalis of a young ram was opened and
the tefticle expofed. The furface of the tefticle was wiped
clean, and a piece of talc was laid upon it. The furface
almoft imediately became more vafcular; five minutes after, the talc was removed and examined in a microfcope,
but no globules could be obferved, only a moifture which
appeared to be ferum. Ten minutes after, there were irregular maffes formed on the talc, fome tranfparent, with
determined edges, but no globules: fifteen minutes after,
nearly the fame.

At twenty minutes, there was an appearance of globules.

At twenty-five minutes, there were globules in clufters;
but I could not fay exactly what thofe globules were.

At thirty-five minutes, the globules more diftinct, more
diffufed, and numerous.

At fifty-five minutes, the globules ftill more perfect and
diftinct.

At feventy, the globules more irregular, and of courfe
lefs diftinct.

At eighty-five, the globules more diftinct and numerous.

At one hundred, more irregular and lefs diftinct, form-ing little maffes.

At two hours, the maffes more tranfparent, and the glo-bules fewer.

At two hours and an half, the maffes tranfparent, and no diftinct globules.

At four hours, fome tranfparent maffes appearing to contain globules

At feven hours, diftinct globules and numerous.

At eight hours, the globules more dictinct and fome-what larger.

At nine hours, lefs appearance of globules.

At twenty-one hours, the tefticle was covered with lint, and the fkin brought over and kept together with a ligature, and allowed to remain for twelve hours, which, from the firft, was thirty-three hours ; when it was opened, it was wiped dry, and a piece of talc applied for five minutes ; the quantity of fluid very fmall, but containing globules fmall and numerous.

N. B. In this time when the tefticle was covered, there were ftrong adhefions took place between the tefticles and tunica vaginalis, which fhows that probably the inflammation moved back to the adhefive ftage whenever two fimilar fur-faces were oppofed.

Forty hours, the above repeated and the globules a lit-tle more diftinct.

Forty-four hours, the appearance of globules very di-ftinct, and it looked like common matter diluted.

EXPERIMENT II.

An opening was made through the linea alba below the navel, feveral inches long, into the cavity of the belly of a dog, care being taken that no blood fhould pafs into that ca-vity ; a piece of talc was applied to the peritoneum fo as to be covered with the fluid which lubricates that furface ; to do which, it was found neceffary to draw it over fome confiderable furface : this fluid was examined in the field of the microfcope, and appeared to contain fmall femitranf-parent globules, few in number, fwimming in a fluid.

The lubricating fluid in the cavity of the abdomen, ap-pears from repeated experiments on healthy dogs, to be fo

fmall in quantity as only to give a polifh to the different furfaces, but not fufficient to have a drop collected.

After five minutes, the furfaces had more moifture upon
them, which being examined as before, the globular appearance was more diftinct.

In fifteen minutes, the furfaces were more vafcular; a
portion of inteftine was wiped dry, and a piece of talc applied to it; the fluid collected on it had a great number of
globules, which were fmaller than thofe at firft obferved.

In an hour, this portion of inteftine had its blood-veffels
confiderably increafed in number; the whole furface appearing of an uniformly red colour: this was wiped dry
and a piece of talc laid upon it; the fluid collected did not
appear to be made up of globules, but of very fmall parts
which had fome tranfparency, but not exactly regular in
their figure, which became ftill more evident on drying,
when they loft the tranfparency altogether; thefe were
moft probably coagulating lymph.

This was repeated upon the furface of the fpleen, which
had its furface exceffively red, from the increafed number
of fmall veffels carrying red blood, and the refult was exactly fimilar.

From thefe experiments, the fluid which lubricates the
peritoneum feems to undergo changes, in confequence of
expofure, and at laft, when inflammation takes place, to
have coagulating lymph fubftituted for it.

Although the lubricating fluid of the peritoneum is fo
fmall in quantity in a natural ftate, yet before that cavity
has been expofed for half an hour, the quantity is much increafed, and has a motled appearance of oil and water; but
from the appearance in the microfcope, it is only an increafe of the original fluid with fome coagulating lymph,
although miftaken by fome anatomifts for an oily lubricating liquor.

EXPERIMENT III.

At half paft feven o'clock in the morning, an incifion was
made with a lancet into the upper flefhy part of a young
ram's thigh, into which was introduced a filver canula, a-
bout a quarter of an inch in diameter, and three quarters
long, with a great number of fmall holes in the fides, and
open at the bottom; it was faftened by means of ligatures
to the fkin, and a fmall cork adapted to it.

The blood was fponged out feveral times, and the cork kept in during the intervals. At half paft nine the cork was withdrawn, and the canula was found to contain a fluid ; a piece of talc was dipped in it, and the appearance was evidently globular, exactly like the red globules without the colour.

At eleven; the quantity of fluid much increafed, and the fame appearance.

At one, the quantity half filling the pipe, of a reddifh brown colour ; the globules more numerous, without colour when diluted with water.

At three, the quantity confiderable, the globules fmaller, freer from colour.

At half paft five, the fame.

EXPERIMENT IV.

In the fame manner the canula was introduced into the flefhy part of an afs's thigh, at nine in the morning ; and at one o'clock, as alfo at two, there was a fluid tinged with red globules.

At four, there were no diffufed globules, but there appeared to be fmall flakes in a tranfparent fluid ; however they proved to be clufters of globules.

At feven, next morning, which was twenty-two hours, there was found in the canula, common pus.

From the experiments on internal furfaces, it would appear that pus was formed coeval with its fecretion ; but from Mr. Home's experiments, page 51, it would rather appear that the globules were not formed till fome time after fecretion, and this fooner or later, according to circumftances, which we probably do not know.

So far thefe experiments explain the progrefs of fuppu- ration on internal furfaces, and I fhall now give its progrefs on the cutis, when deprived of its cuticle, from Mr. Home's Differtation on that fubject beforementioned.

" I applied a blittering plafter of the fize of a half-crown piece to the pit of the ftomach of a healthy young man. In eight hours a blifter arofe, which was opened, and the contents removed ; they were fluid, tranfparent, and coagulated by heat; had no appearance of globules, when examined by the microfcope; and in every refpect refembled the ferum of blood. The cuticle was not removed, but allowed to collapfe, and the fluid, which was formed upon the furface of the cutis, was examined from time to time by a microfcope, to dtermine as accurately as poffible the changes which took place.

" The better to do this, as the quantity in the intervals ftated below muft be exceedingly fmall, a piece of talc, very thin and tranfparent, was applied to the whole furface, and covered with an adhefive plafter; and the furface of the talc applied to the fkin was removed and examined by the microfcope, applying a frefh piece of talc after every examination, to prevent any miftake which might have a-rifen from the furface not being quite clean.

" The fluid was examined by the microfcope, to afcer-tain its appearance ; but as the aqueous part in which the globules of pus fwim, is found by experiment to coagulate, by adding to it a ftaturated folution of fal ammoniac, which is not the cafe with the ferum of the blood nor the tranfpa-rent part of the milk, I confidered this as a property peculiar to pus ; and confequently that it would be a very good teft by which to afcertain the prefence of true pus.

" In eight hours.—From the time the blifter was applied, the fluid difcharged was perfectly tranfparent, and did not coagulate with the folution of fal ammoniac.

" Nine hours.—The difcharge was lefs tranfparent; but free from the appearance of globules.

" Ten hours.—The difcharge contained globules which were very fmall, and few in number.

" Eleven hours.—The globules were numerous, but ftill the fluid did not coagulate with the folution of fal am-moniac.

" Twelve hours.—The appearance much the fame as be-fore.

" Fourteen hours.—The globules a little larger, and the fluid appeared to be thickened by a folution of fal am-moniac.

" Sixteen hours.—The globules feemed to form themfelves into maffes ; but were tranfparent.

" Twenty hours.—The globules were double the fize of thofe firft obferved at ten hours, and gave the appear-ance of true pus, in a diluted ftate: the fluid was coagu-lated by a folution of fal ammoniac ; the globules at the fame time remaining perfectly diftinct, fo that I fhould confider this as true pus.

" Twenty-two hours.—No change appeared to have taken place.

" Thirty-two hours.—The fluid was confiderably thicker
in confiftence, the number of globules being very much
increafed : but in no other refpect that I could obferve,
did it differ from that formed twenty hours after the application of the blifter."

To afcertain the progrefs of fuppuration on canals, or fecreting furfaces, I have often examined the matter on a
bougie, that had been introduced into the urethra, and
found it to be formed earlier than either of the times beforementioned ; Mr. Home's experiment makes it five
hours ; but we often find a gonorrhœa coming on at once,
not having in the leaft been preceded by a leading difcharge.

Since that period experiments have been made on pus,
from different kinds of fores, with an intention to afcertain the nature of the fore by the refult of fuch analyfis.
That fores give very different kinds of pus is evident to the
naked eye, and that the different parts of which the blood
is compofed will come away in different proportions,
we can make no doubt ; and we find that whatever is in
folution in the blood, comes away more in one kind of pus
than another, which are all fo many deviations from true
pus ; we may alfo obferve, that fuch kinds of pus change,
after being fecreted, much fooner than true pus, which
will be obferved by and by. From all this I fhould be apt
to conceive, that fuch experiments will throw little light on
the fpecific nature of the difeafe, which is the thing wanted. From fuch experiments we may find out that pus,
from a venereal bubo in its height of malady, or that from
a cancer, is bad matter, but cannot afcertain the difference
between thofe two matters and all others, nor the fpecific
difference between the two. The fmall-pox, although as
malignant a difeafe as any, and one which produces a pus
as replete with poifonous particles as any, yet gives a true
pus, when not of the confluent kind, which difpofition is
not fmall-pox. The reafon why it is good pus, is, becaufe
its inflammation is of the true fuppurative kind ; and the
reafon why it is of the true fuppurative kind, is, becaufe the
parts have the power of curing themfelves, juft as much as
in any accident which happens to fuch a conftitution ; but
this is not the cafe with either the venereal difeafe or the
cancer; from the moment thefe fet out, their difpofitions
tend to become worfe and worfe ; but the venereal bubo, if
mercury is given fo as to affect it, foon gives us another

kind of pus, although this has the poifon equally in it;
therefore it is not the circumftance of containing a poifon
which makes it what is called a bad pus, but its being formed from a fore that has no difpofition to heal : as we cannot
give healing action to a cancer, fo we never can have a
good pus. The obfervation refpecting the fmall-pox is applicable to the venereal gonorrhœa ; for this complaint having the power of curing itfelf, its pus is good in proportion
to that power ; but as the periods of cure are not fo determined as in the fmall-pox, neither is its time in producing good pus fo determined ; but like the fmall-pox, as
well as the venereal difeafe when it is healing, we have
good pus, although it contains the poifon.

From the above experiments it muft appear unneceffary
to give the chemical analyfes of what is commonly called
pus, for whatever comes from a fore has that name, although
very different in many cafes from what I fhould call true pus;
and we fhall find in thofe fores that have fome fpecific
quality which hinders them from healing, that the difcharge
is not pus. Probably the chemical properties may be
nearly the fame in them all.

II. OF THE PROPERTIES OF PUS.

Pus, in the moft perfect ftate, has at the firft view certain
peculiar qualities. Thefe are principally colour and confiftence ; but it appears that the colour takes its rife from
the largeft portion of the whole mafs being compofed of
very fmall round bodies, very much like thofe fmall round
globules, which fwimming in a fluid make cream : I fhould
fuppofe thofe round globules to be white in themfelves, as
cream would appear to be ; although it is not neceffary that
the fubftance of matter which reflects a white, fhould be
itfelf white ; for a vaft number of tranfparent bodies being
brought together will produce a white, fuch as broken glafs,
broken ice, water covering globules of air, making froth, etc.

These globules swim in a fluid, which we should at first suppose to be the serum of the blood, for it coagulates with heat like serum, and most probably is mixed with a small quantity of coagulating lymph ; for pus in part coagulates, after having been discharged from the secreting vessels, as mucus is observed to do. But although it is thus far similar to serum, yet it has properties that serum has not. Observing there was a similarity between pus and milk, I tried if the fluid part of pus could be coagulated with the juice of the stomach of other animals, but found it could not. I then tried it with several mixtures, principally with the neutral salts, and found that a solution of sal ammoniac coagulated this fluid ; not finding that a solution of this salt coagulated any other of our natural juices, I concluded that globules swimming in a fluid that was coagulable by this salt was to be considered as pus, and would be always formed in sores that had no peculiar backwardness to heal.

The proportion that these white globules in the pus bear to the other parts depends on the health of the parts which formed it ; for when they are in a large proportion, the matter is thicker and whiter, and is called good matter ; the meaning of which is, that the solids which produced it are in good health ; for these appearances in the matter are no more than the result of certain salutary processes going on in the solids, the effect of which processes is, to produce the disposition on which both suppuration and granulation depend ; all this is a good deal similar to the formation of milk ; for in the commencement of the secretion of this fluid, it is at first principally serum, and, as the animal advances towards delivery, the globules are forming and become more in quantity, and the animal that has them in largest quantity has the richest milk ; likewise when they are naturally leaving off secreting milk, it again takes an exact retrograde motion ; and we may also observe, that if any local affection attacks this gland, such as inflammation, the milk is falling back to the state I have been now describing ; or if any constitutional affection takes place, such as fever, etc. then this gland suffers in the same manner.

Pus is specifically heavier than water ; it is probably nearly of the same weight with blood or any other animal substance rendered fluid.

Pus, befides the abovementioned properties, has a fweetifh and maukifh tafte, probably from having fugar in it,
which is very different from moft other fecretions, and the
fame tafte takes place, whether it is pus from a fore, viz. an
ulcer, or an irritated inflamed furface. Thus, if any have
an ulcer in their nofe, mouth, throat, lungs, or parts adjacent, fo that the matter fhall come into the mouth unaltered by putrefaction, they will be able to tafte it from
its having this property ; whereas the mucus and faliva of
thofe parts is taftelefs. The fame thing happens when an
irritation to inflammation takes place on the furface of thofe
parts without ulceration.

If the internal furface of the nofe is inflamed, fo that
when we blow it on a white handkerchief, we fee the fubftance difcharged of a yellow colour ; we alfo find that
when we draw up the fame fubftance into the mouth, that
it has a fweetifh maukifh tafte. If it is the furface of the
mouth or throat that difcharge this matter, the fame tafte
is obfervable ; and if it is brought up from the trachea and
lungs, in confequence of the common effects of a cold on
thofe parts, the fame tafte is alfo to be obferved ; fo that
pus, from whatever furface, whether an irritated natural
furface, or the furface of a common fore, has this property.

Pus has a fmell in fome degree peculiar to itfelf ; but this
differs ; fome difeafes, fuch for inftance, as the venereal
gonorrhoea, it is pretended may be known by the fmell.

To afcertain the properties of pus, or to diftinguifh it
from mucus, it has, with mucus, been put to the teft of
chemiftry. Solution in menftrua, and precipitation were
thought to be a teft of their diftinction.

This principle in its very firft appearance is unphilofo-
phical, and was at the very firft treated by me as abfurd.
I conceived that all animal fubftance whatever, when in folution, either in acids or alkalies, would then be in the fame
ftate, and therefore that the precipitation would be the fame
in all. Calcarious earth, when diffolved in an acid, (for
inftance, the muriatic) is in that acid in the fame ftate,
whether it has been diffolved from chalk, lime-ftone, mar-
ble, or calcarious fpar ; and the precipitations from all are
the fame.

However, whatever my opinion might be, yet bold af-
fertions, the refult of defcribed experiments, made me a-

void falling into the fame error, of defcribing what I never had feen ; I made, therefore, fome experiments on this
fubject ; and in confequence of having previoufly formed
the abovementioned opinion, I was more general in my experiments. I made them on organic animal matter, as
well as on inorganic, and the refult was the fame in all.

As organic animal matter, I took mufcle, tendon, car-
tilage, gland, viz. liver and brain.

As inorganic animal matter, I took pus, and the white
of an egg, and diffolved each in the vitriolic acid, and then
precipitated the folution with vegitable alkali.

Each precipitation I examined with fuch magnifiers as
plainly fhewed the forms of the precipitate ; all of which
appeared to be fleaky fubftances.

The precipitate by the volatile alkali, appeared exactly
the fame.

To carry thofe experiments a little further, I diffolved
the fame fubftance in the vegitable cauftic alkali, and precipitated the folution with the muriatic acid, and examined each precipitate with the microfcope, and the appearance was the fame, viz. a fleaky fubftance, without any
regular form.

To fee how far the nature of fores, might be afcertained from the nature of their difcharge, matter from a cancerous fore has been analyfed, and the refult has been, that
fuch matter differs from true pus ; but this explains nothing more than what the naked eye can perceive, that it is
not pus ; but it will not fhew the fpecific difference between the matter from a cancer and matter from a venereal bubo, where mercury has not been given, nor will it tell
that one is cancer and the other is venereal. We might as
well analyfe the urine at different times, in order to afcertain the nature of kidnies at thofe times.

The quality of pus is always according to the nature of
the parts which produce it ; and whatever fpecific qualities the parts may have befides, the pus has alfo this fpecific quality ; hence we have venereal matter from venereal
fores, fmall-pox matter from fmall-pox fores, cancerous
matter from cancerous fores, etc. It is not in the leaft affected by the conftitution, except the parts which produce
it are alfo affected by the conftitution.

VOL. II. U

Pus is fo far of the fame fpecific nature with the part which produces it, that it does not become an irritator to that part ; it is perfectly in harmony with it, the part is not in the leaft fenfible of it ; therefore the pus of a fuppurating furface is not an irritator to the fame furface, but may be an irritator to any other not of the fame kind; hence no fuppurating furface of any fpecific kind can be kept up by its own matter, for if this had not been the cafe, no fore of any fpecific quality, or producing matter of an irritating quality, could ever have been healed. This is fimilar to every other fecretion of ftimulating fluids, as the bile, tears, etc. for thofe do not ftimulate their own glands or ducts, but are capable of ftimulating any other part of the body. The venereal gonorrhœa, fmall-pox, etc. healing or recovering of themfelves, are ftriking inftances of this ; howevever, we find matter under certain circumftances ftimulating its own fore, and alfo fecretion ftimulating their own canals, as the fecretions of the inteftines ftimulating themfelves ; but how far this may not arife from one part of the inteftines being fo difeafed, as to fecrete a ftimulating fluid, and coming to a found part, ftimulates that only I will not determine. This I am certain happens to the rectum and anus ; for it very often happens in purging, that the watery ftools fhall irritate thofe parts fo much as to make them feel as if they were fcalded. This idea feems reafonable on another principle ; for when we confider matter in the grofs, we fhall find that it is often mixed with extraneous fubftances which make no part of it, being probably ftrained from the blood ; and alfo, probably, undergoing a change afterwards from its not being pure pus; nor do thefe always arife entirely from the nature of the fore ; for they are produced by fores of very different fpecific qualities, it being the fpecies of matter itfelf which arifes from the nature of the fore ; however, the kind of fore will often produce more or lefs of this extraneous matter, and this additional fubftance may act as a ftimulus on every kind of fore.

What I have confidered thus far, is the natural procefs of a found conftitution and found parts ; fince a fore that is going through all the natural ftages to a cure, is not to be called a difeafe.

A proof of this is, that whenever a real difeafe attacks either the fuppurating furface, or the conftitution, thefe proceffes of nature are deftroyed, and the very reverfe takes place; the production of true pus ceafes, and the fluid becomes changed in fome meafure in proportion to thefe morbid alterations; in general it becomes thinner and more tranfparent, as if the part was returning back to the adhefive ftate, it partakes more of the nature of the blood, as is the cafe in moft other fecretions under fimilar circumftances. This, in common language, is not called pus, but fanies.

Pus, arifing from fuch ftate of fores, has more of the ferum, and frequently of the coagulating lymph in it, and lefs of the combination that renders it coagulable with a folution of fal ammoniac. It has a greater proportion alfo of the extraneous parts of the blood that are foluble in water, fuch as falts; and becomes fooner putrid. The two laft fpecies of matter not being of the fame fpecific nature with the fore, they have the power of ftimulating even their own fore.

On this laft account too, pus becomes more irritating to the adjoining parts, with which it comes in contact, producing excoriation of the fkin, and the ulcerative inflammation; as the tears, when they run out, excoriate the fkin of the cheek from the quantity of falts which they contain. From this effect the matter has been called corrofive, a quality which it has not; the only quality which it poffeffes being that of irritating the parts with wich it comes in contact, in fuch a manner that they are removed out of the way by the abforbents, as will be defcribed when treating on ulceration.

In thefe inftances of the change in the pus, we may fay that the change is effected by the decompofition and new combination not being carried on fo perfectly; this may probably depend on the fecreting veffels having loft their due ftructure and action, and this appears to be fo much the cafe, that they not only fail in this operation, but the other offices of thofe veffels, viz. the production of granulations is alfo checked; for the veffels forming themfelves into a certain ftructure which fits them for fecreting pus, it is fo ordered, that the fame ftructure alfo fits them for producing granulations, and thus thofe two proceffes are concomitant effects of the fame caufe, which caufe is a peculiar organization fuperadded to the veffels of the part,

What organization this may be is not in the least known, nor muſt we wonder at this, for it is exactly the ſame with every other organ of ſecretion, about all which we are e-qually ignorant; indeed, ſome of the differences between one gland and another are made out, and alſo ſomething of their general ſtructure; but not in ſuch a way as can lead us to the actions and operations of the ſeveral parts upon which the nature of the different ſecretions depend, ſo as to enable us to conclude à priori, that this or that gland muſt ſecrete this or that peculiar juice.

Pus, from ſeveral circumſtances often attending it, would appear in general to have a greater tendency to pu-trefaction than the natural juices have ; but I very much ſuſpect that this is not really the caſe with pure pus ; for when it is firſt diſcharged from an abſceſs, it is in general perfectly ſweet. There are, however, ſome exceptions to this, but theſe depend on circumſtances entirely foreign to the nature of pus itſelf. Thus, if the abſceſs had any communication with the air while the matter was confined in it, (as is frequently the caſe with thoſe in the neighbour-hood of the lungs) or if it has been ſo near the colon or rec-tum, as to have been infected by the fœces, under ſuch circumſtances we cannot wonder that it becomes putrid : matter formed early in the ſtate of ſuppuration, either in ab-ſceſſes, or more eſpecially in conſequence of any external violence committed on the ſolids, has always in it a por-tion of blood ; or if ſome parts of the ſolids mortify and ſlough, theſe will mix with the matter ; the ſame thing happens when the inflammation has ſomething of the eryſi-pelatous diſpoſition, ſo as to have produced a mortification in the ſeat of the abſceſs ; in all ſuch circumſtances we find the pus has a greater tendency to putrify than the pure or true pus, which comes to be diſcharged afterwards in ſound abſceſſes or healing ſores ; and accordingly the matter from recent ſores becomes very putrid between every dreſſing ; whereas, when the ſame ſores are further advanced, it is perfectly ſweet at the ſame periods; but although the im-perfect or heterogeneous matter that is formed at firſt is li-able to putrify when expoſed, yet if it is perfectly confined in an abſceſs, it will remain a conſiderable time without pu-trefaction ; the ſuppuration, however, in conſequence of the eryſipelatous inflammation, which is often attended with ſuppuration produced by internal mortification, is, as we have obferved, an exception to this rule ; for although confined from external air, yet the matter becomes foon putrid, and this moft probably arifes from the folids themfelves firft becoming putrid.

A fimilar obfervation may be made with refpect to fores which have been in the habit of difcharging good pus ; for if by any accident an extravafation of blood is produced in thefe parts, or a difpofition is brought on to throw out blood, which mixes with the pus, the difcharge changes from its former fweetnefs, and becomes much more putrid and offenfive. It appears that pure matter, although eafily rendered fufceptible of change by extraneous additions, is in its own nature pretty uniform and immutable. It appears fo unchangeable, that we find it retained in an abfcefs for weeks, without having undergone any change ; but thefe qualities being only to perfect pus ; for if a fore from a found ftate changes its difpofition and becomes inflamed, the matter now produced from it, though there be no extravafated blood or dead folids, becomes much fooner putrid than that which was difcharged before this alteration of difpofition, and fhall become much more irritating, as has already been obferved.

From the abovementioned confiderations, we can explain why the difcharge in many fpecific difeafes, although not in all, is fo much more offenfive than in common fores ; for in thefe cafes it is commonly not true pus, and is generally mixed with blood.

In the fame manner, likewife, where there are difeafed bones, or other extraneous bodies which excite irritation, fometimes even to fo great a degree as to caufe the veffels to bleed, and often wounding the veffels of the part, the matter is always found to be very offenfive, one mark (although not commonly accounted for) of a difeafed bone.

Our filver probes are rendered almoft black, when introduced into the difcharge of an unhealthy fore ; preparation of lead are the fame, when applied to fuch matter. It even diffolves animal fubftance ; if, for inftance, a frefh wound has its lips brought together and held there with fticking plafter fpread upon leather, we fhall find if the wound fuppurates, that the parts of the ftraps of leather going over the wound will be between the firft and fecond dreffing quite diffolved, dividing the ftraps into their two ends ; and the plafter, which commonly has lead in it, ſhall become black, where it has come in contact with this matter. This change in the colour of metals is alſo produced by eggs, when not perfectly freſh, although not become putrid ; and probably this property is affiſted by the boiling or roaſting. Dr Crawford, in his experiments on the matter of cancers and animal hepatic air, attributes the diſſolution of the metals to that air†.

III. OF THE USE OF PUS.

The final intention of this ſecretion of matter is, I believe, not yet underſtood, although almoſt every one thinks himſelf able to aſſign one ; and various are the uſes attributed to it. It is by ſome ſuppoſed to carry off humors from the conſtitution. It is ſometimes ſuppoſed a conſtitional diſeaſe changed into a local one, and ſo diſcharged or thrown out of the body, either in form of, or with the pus, as in thoſe caſes to be called critical abſceſſes ; but even thoſe who ſee this final intention are very ready to overturn it, by ſuppoſing that this matter is capable of being taken back again into the conſtitution by abſorption, and producing much worſe evils than thoſe it was meant to relieve. I believe that the ſuppoſed caſes of abſorption are more numerous than thoſe where it is ſuppoſed to relieve ; if ſo, then by their own account nothing is gained. Or it is preſumed to carry off local complaints from other parts of the body by way of derivation, or revulſion ; for this reaſon ſores, as iſſues, are made in ſound parts, to allow other ſores to be dried up ; or even with a view to oblige parts to diſſolve themſelves into pus, as indurated ſwellings; but we have endeavoured to ſhow that the ſolids make no part of pus.

A ſecretion of pus is alſo looked upon as a general prevention of many, or of all the cauſes of diſeaſe ; iſſues, therefore, are made to keep off both univerſal diſeaſe as well as local. But I am apt to believe that we are not yet

† Philoſ. Tranſact. vol. 8oth, year 1790, part 2nd. page 385.

well, or perhaps at all acquainted with its ufe, for it is common to all fores ; takes place in the moft perfect degree in thofe fores which may be faid to be the moft healthy, and efpecially in thofe where the conftitution is moft healthy.

We find alfo that very large difcharges, when proceeding from a part which is not effential to life, produces very little change in the conftitution, and as little upon being healed up, whatever fome people may fuppofe to the contrary.

One might naturally imagine, that it was of fervice to the fore which formed it, to keep it moift, etc. for all internal furfaces have their peculiar moifture ; but as a fore is to heal, and if allowed to dry, fo foon as to form a fcab, then a fore is difpofed to form no more pus, and heal fafter ; it is the mode of dreffing external fores that keeps up this fecretion, which in this refpect maintains the fore in the ftate of an internal one ; but this will not account for the formation of an abfcefs, which is the formation of pus we can beft account for, fince it produces the expofure of internal furfaces ; in many cafes it is of fingular fervice, to procure the fecond mode of cure, and open a communication between the difeafe and the external furface of the body.

It alfo forms a paffage for the exit of extraneous bodies ; but all thefe are only fecondary ufes.

CHAPTER VI.

THE ULCERATIVE INFLAMMATION.

I N confidering the origin and courfe of the blood, it would have been moft natural to have confidered ab-forption, or the abforbing veffels ; for in one point of view, they may be confidered as the animal confifting of fo many mouths, every thing elfe depending upon them, or belonging to them ; for in tracing thefe dependences we find that there exifts ultimately little elfe but abforbents. The ftomach and the organs connected with it in fuch animals as have a ftomach, are to be confidered as fubfer-vient to this fyftem ; and many an animal is to be confi-dered as confifting of a number of ftomachs ; a piece of co-ral, for inftance, appears to be no more than a thoufand ftomachs, all taking in food for digeftion, and abforption for increafe, and fupport of the whole ; for each ftomach does not increafe, as the piece of coral increafes, but they multiply in number, and of courfe the whole piece of coral increafes ; for although each appears to be a diftinct animal, yet it is not fo; but as this is too general a view of this fyftem for our prefent purpofe, I fhall leave it, and confine myfelf principally to the ufes of the abforbents in the difeafes of which I am going to treat ; and as one of their ufes in difeafes, and indeed the principal one, has not been defcribed, nor in-deed in the leaft conjectured, that it may be clearly under-ftood or diftinguifhed from the other known ufes, I fhall relate firft the more common ufes which have been for-merly affigned to this fyftem.

Firft, the abforbents take up extraneous matter, in which is included nourifhment.

Secondly, fuperfluous and extravafated matter, whether natural or difeafed.

Thirdly, the fat.

Fourthly, they produce a wafte of parts, in confequence of which mufcles become fmaller, bones become lighter, etc. Although thefe two laft effects were perhaps not expreflly faid to be carried on by abforption, either by veins, or any other fyftem of veffels, yet we muft fuppofe they were underftood : fo far the abforbents have in general been confidered as active parts in the animal œconomy ; but from a further knowledge of thefe veffels, we fhall find that they are of much more confequence in the body than has been imagined, and that they are often taking down what the arteries had formerly built up ; removing whole organs, becoming modellers of the form of the body while growing ; alfo removing many difeafed and dead parts, which were beyond the power of cure ; of all which I fhall now take particular notice.

As thefe veffels are productive of a vaft variety of effects in the animal œconomy, which are very diffimilar in the intention and effect, they may be reviewed in a variety of lights, and admit of a variety of divifions. I fhall confider them in two views : firft, as they abforb matter, which is not any part of the machine ; fecondly, as they abforb the machine itfelf.

The firft of thefe is the well-known ufe, the abforption of matter, which is no part of the machine. This is of two kinds, one exterior matter, in which may be ranked every thing applied to the fkin, as alfo the chyle ; and the other interior, fuch as many of the fecreted juices, the fat, and the earth of bones, etc*. Thefe are principally with a view to its nourifhment, and alfo anfwer many other purpofes; fo that the action of abforbing foreign matter is extremely extenfive ; for befides its falutary effects, it is often the caufe of a thoufand difeafes, efpecially from poifons, none of which are to my prefent purpofe.

In the fecond of thefe views, we are to confider them as removing parts of the body itfelf, in which they may be

---

* It may be neceffary to remark here, that I do not confider either the fat, or the earth of bones, as a part of the animal; they are not animal matter ; they have no action within themfelves. They have not the principle of life.

viewed in two lights. The firſt is, where only a waſting is produced in the whole machine, or part, ſuch as in the waſting of the whole body, from an atrophy; or in a part, as in the waſting of the muſcles of the leg, etc, from ſome injury done to ſome nerve, tendinous part, or joint; all of which I call interſtitial abſorption, becauſe it is removing parts of the body out of the interſtices of that part which remains, leaving the part ſtill as a perfect whole*. But this mode is often carried further than ſimply waſting of the part; it is often continued till not a veſtage is left, ſuch as the total decay of a teſticle, ſo that the interſtitial abſorption might be underſtood in two ſenſes.

The ſecond is, where they are removing whole parts of the body. This may be divided into the natural, and the diſeaſed†.

In the natural they are to be conſidered as the modellers of the original conſtruction of the body; and if we were to conſider them fully in this view, we ſhould find that no alteration can take place in the original formation of many of the parts, either in the natural growth, or that formation ariſing from diſeaſe, in which the abſorbents are not in action, and take not a conſiderable part: this abſorption I ſhall call modelling abſorption. If I were to conſider their powers in this light, it would lead me into a vaſt variety of effects, as extenſive as any principle in the animal œconomy, for a bone cannot be formed without it, nor probably many other parts. A part which was of uſe in one ſtage of life, but which becomes entirely uſeleſs in another, is thus removed. This is evident in many animals; the thymus gland is removed; the ductus arterioſus, and the membrana pupilaris is removed. This proceſs is, perhaps, more remarkable in the changes of the inſect, than in any other known animal. Abſorption in conſequence of diſeaſe, is the power of removing complete parts of the body, and is in its operation ſomewhat ſimilar to the firſt of this diviſion, or modelling proceſs, but very different in the intention, and therefore in its ultimate effects.

---

* This mode of abſorption has always been allowed, or ſuppoſed, whether performed by the lymphatic veins, or lymphatics.

† Theſe uſes I claim as my own diſcovery. I have taught them publicly ever ſince the year 1772.

This procefs of removing whole parts in confequence of
difeafe, in fome cafes produces effects which are not fimilar to one another ; one of thefe is a fore or ulcer, and I
therefore call it ulcerative.   In other cafes no ulcer is produced, although whole parts are removed, and for this
I have not been able to find a term ; but both may be denominated progreffive abforption.

This procefs of the removal of a whole folid, part of the body, or that power which the animal œconomy has taking part
of itfelf into the circulation by means of the abforbing veffels,
whenever it is neceffary, is a fact that has not in the leaft
been attended to, nor was it even fuppofed, and having
now been noticed, I mean to give a general idea of it.   I
may juft be allowed once more to obferve, that the oil, or
fat, of animals, and the earth of bones, have always been
confidered as fubject to abforption ; and fome other parts
of the body being liable to wafting, have been fuppofed to
fuffer this by abforption ; but that any folid part fhould totally be abforbed, is a new doctrine.

This ufe of the abforbents I have long been able to de-
monftrate ; and the firft hints I received of it, were in the
wafte of the fockets of the teeth, as alfo in the fangs of
the fhedding teeth.

It may be difficult at firft to conceive how a part of
the body can be removed by itfelf ; but it is juft as difficult
to conceive how a body can form itfelf, which we fee daily
taking place ; they are both equally facts, and the knowledge of their mode of action, would anfwer perhaps very little purpofe ; but this I may affert, that whenever any
folid part of our bodies undergoes a diminution, or is broken in upon, in confequence of any difeafe, it is the abforbing fyftem which does it.

When it becomes neceffary that fome whole living part fhould
be removed, it is evident that nature, in order to effect this,
muft not only confer a new activity on the abforbents,
but muft throw the part to be abforbed into fuch a ftate as
to yield to this operation.

This is the only animal power capable of producing fuch
effects, and like all other operations of the machine arifes
from a ftimulus, or an irritation ; all other methods of deftruction being either mechanical or chemical.   The firft
by cutting inftruments, as knives, faws, etc. the fecond by
cauftics, metalic falts, etc.

The procefs of ulceration is of the fame general nature in all cafes ; but fome of the caufes and effects are very different from one another.

The knowledge of the ufe of this fyftem is but of late date ; and the knowledge of its different modes of action is ftill later. Phyfiologifts have laboured to account for its modes of action ; and the principle of capillary tubes was at firft the moft general idea, becaufe it was a familiar one. But this is too confined a principle of an animal machine, nor will it account for every kind of abforption. Capillary tubes can only attract fluids ; but as thefe inquirers found that folids were often abforbed, fuch as fchirrous tumors, coagulated blood, the earth of bones, etc. they were driven to the neceffity of fuppofing a folvent ; this may or may not be 'true ; it is one of thofe hypothefes that can never be proved or difproved, and may for ever reft upon opinion. But my conception of this matter is, that nature leaves as little as poffible to chance, and that the whole operation of abforption is performed by an action in the mouths of the abforbents : but even under the idea of capillary tubes, phyfiologifts were ftill obliged to have recourfe to the action of thofe veffels to carry it along after it was abforbed ; and might therefore as well have carried this action to the mouths of thefe veffels.

As we know nothing of the mode of action of the mouths of thefe veffels, it is impoffible we can form any opinion that can be relied upon ; but as they are capable of abforbing fubftances in two different ftates, that of folidity and fluidity, it is reafonable to fuppofe that they have different modes of action ; for altho ugh any conftruction of parts that is capable of abforbing a folid, may alfo be fuch as is capable of abforbing a fluid ; yet I can fuppofe a conftruction only capable of abforbing a fluid, and not at all fitted for abforbing a folid, though this is not likely ; and to fee the propriety of this remark more forcibily, let us only confider the mouths of different animals, and I will venture to fay, that the mouths of all the different animals have not a greater variety of fubftances to work upon, than the abforbents have, and we may obferve that with all the variety of mouths in different animals, this variety is only for the purpofe of adapting them to abforb folids, which admit of great variety in form, texture, etc. every one being capable of abforbing fluid matter, which admits of no variety.

This procefs of the removal of parts of the body, either by interftitial or progreffive abforption, anfwers very material purpofes in the machine, without which many local difeafes could not be removed, and which, if allowed to remain, would deftroy the perfon.   It may be called in fuch cafes, the natural furgeon.

It is by the progreffive abforption, that matter or pus, and extraneous bodies of all kinds; whether in confequence of or producing inflammation and fuppuration are brought to the external furface ; it is by means of this that bones exfoliate ; it is this operation which feparate floughs ; it is the abforbents which are removing whole bones, while the arteries are fupplying new ones ; and although in thefe laft cafes of bones it arifes from difeafe, yet it is fomewhat fimilar to the modelling procefs of this fyftem in the natural formation of bone ; it is this operation that removes ufelefs parts, as the alveolar proceffes, when the teeth drop out,or when they are removed by art ; as alfo the fangs of the fhedding teeth, which allows them to drop off ; and it is by thefe means ulcers are formed.

It becomes a fubftitute in many cafes for mortification, which is another mode of the lofs of fubftance ; and in fuch cafes it feems to owe its taking place of mortification to a degree of ftrength or vigor, fuperior to that where mortification takes place ; for although it arifes often from weaknefs, yet it is an action,while mortification is the lofs of all action.   In many cafes it finifhes what mortification had begun, by feparating the mortified part.

Thefe two modes of abforption, the interftitial and the progreffive, are often wifely united, or perform their purpofes often in the fame part which is to be removed ; and this may be called the mixed, which I believe takes place in moft cafes, as in that of extraneous bodies of all kinds coming to the fkin ; alfo in abfceffes, when in foft parts.   It is the fecond kind of interftitial abforption, the progreffive and the mixed, that becomes moftly the object of furgery, although the firft of the interftitial fometimes takes place, fo as to be worthy of attention.

This operation of the abforption of whole parts, like many other proceffes in the animal œconomy, arifing from difeafe, would often appear to be doing mifchief, by deftroying parts which are of fervice, and where no vifible good appears to arife from it ; for it is this procefs which forms a fore called an ulcer ; fuch as in thofe cafes where the folids are deftroyed upon the external furface, as in old fores in the leg, breaking out anew, or increafing ; but in all cafes it muft ftill be referred to fome neceffary purpofe ; for we may depend upon it, that thofe parts have not the power of maintaining their ground, and it becomes a fubfti-tute for mortification ; and indeed in many ulcers, we fhall fee both ulceration and mortification going on, ulceration removing thofe parts that have power to refift death.

I OF THE REMOTE CAUSE OF THE ABSORP-TION OF THE ANIMAL ITSELF.

THE remote caufe of the removal of parts of the animal appears to be of various kinds, and whatever will produce the following effects, will be a caufe.

The moft fimple intention, or object of nature, feems to be the removal of a ufelefs part, as the thymus gland, membrana pupilaris, ductus arteriofus, the alveoli when the teeth drop, or the cryftaline humor after couching, and pro-bably the wailting of the body from fever either accute or hectic. Thefe parts are removed by the abforbents, either as ufelefs parts or in confequence of ftrength being unne-ceffary while under difeafe, or fuch as not to accord with dif-eafe,*

Another caufe is a weaknefs, or the want of power in the part to fupport itfelf under certain irritations, which may be confidered as the bafis of every caufe of removal of whole parts ; as the abforption of callufes, cicatrices, the gums in falivation ; alfo that arifing from preffure, or irri-

* It might be afked as a queftion, whether the wafte of the conftitution in difeafe arifes from the body becoming ufelefs when under fuch di eafes, as may be obferved of mufcles when their joint, tendon, etc. is difeafed ; or whether it accords bet-ter with the difeafed ftate, and may even tend to a natural cure ?

tating applications, under which may be included the attachment of dead parts to a living one ; all of which may be accounted for upon the fame principle of parts or organs not being able to fupport themfelves under the prefent evil.

From the above account of the final caufe of the abforption of whole parts from difeafe, it would appear that they are capable of being abforbed from five caufes. Firft, from parts being preffed ; fecondly, from parts being confiderably irritated by irritating fubftances ; thirdly, from parts being weakened ; fourthly, from parts being rendered ufelefs ; fifthly, from parts becoming dead. The two firft, for inftance, parts being preffed, and parts being irritated, appear to me to produce the fame irritation ; the third, or weaknefs, an irritation of its own kind ; and the fourth, or parts being rendered ufelefs, and the fifth, or parts becoming dead, may be fomewhat fimilar.

It is probable that every caufe above enumerated is capable of producing every mode ; or rather effect of abforption whether interftitial or progreffive ; but preffure attended with fuppuration always produces the progreffive, whether applied externally or, internally, as in the cafe of abfceffes.

_____

II. OF THE DISPOSITION OF LIVING PARTS TO ABSORB AND TO BE ABSORBED.

The difpofitions of the two parts of the living body, which abforb and are abforbed, muft be of two kinds refpecting the parts ; one paffive and the other active. The firft of thefe is an irritated ftate of the part to be abforbed, which renders it unfit to remain under fuch circumftances ; the action excited by this irritation being incompatible with the natural actions and the exiftence of the parts whatever thefe are, therefore become ready for removal, or yield to it with eafe. The fecond is the abforbents being ftimulated to action by fuch a ftate of parts, fo that both confpire to the fame end.

When the part to be abforbed is a dead part, as nouriſh-ment or extraneous matter of all kinds,then the whole diſ-poſition is in the abſorbents.

When thoſe immediate cauſes ariſe in conſequence of preſſure, it would appear that abſorption takes place more readily under certain circumſtances than others, although the remote cauſes of them appear to be the ſame, therefore ſomething more than ſimple preſſure ; for we find that preſſure from within produces ulceration or abſorption much more readily than from without ; for if it was preſ-ſure only,abſorption then would be according to the quantity of preſſure ; but we find very different effects from the ſame quantity of preſſure under the abovementioned circum-ſtances ; for when from without, preſſure rather ſtimu-lates than irritates ; it ſhall give ſigns of ſtrength, and pro-duce an increaſe of thickening ; but when from within, the ſame quantity of preſſure will produce waſte ; for the firſt effect of the preſſure from without is the diſpoſition to thicken, which is rather an operation of ſtrength ; but if it exceeds the ſtimulus of thickening, then the preſſure be-comes an irritator, and the power appears to give way to it, and abſorption of the parts preſſed takes place, ſo that nature very readily takes on thoſe ſteps which are to get rid of an extraneous body, but appears not only not ready to let extraneous bodies enter the body, but endea-vours to exclude them, by increaſing the thickneſs of the parts.

Many parts of our ſolids are more ſuſceptible of being abſorbed, eſpecially by ulceration, than others, even un-der the ſame or ſimilar circumſtances, while the ſame part ſhall vary its ſuſceptibility according to circumſtances.

The cellular and adipoſe membranes are very particular-ly ſuſceptible of being abſorbed, which is proved by muſ-cles, tendons, ligaments, nerves and blood veſſels being found frequently deprived of their connecting membrane and fat ; eſpecially in abſceſſes, ſo that ulceration often takes a roundabout courſe to get to the ſkin, following the track of the cellular membrane ; and the ſkin itſelf, when the preſſure is from within, is much leſs ſuſceptible of ul-ceration than the cellular and adipoſe membrane, which re-tards the progreſs of abſceſſes, when they are ſo far advanc-ed, and alſo becomes the cauſe of the ſkin's hanging over spreading ulcers, which are spreading from the same cause more especially too, if the part ulcerating is an original part. Ulceration never takes place on investing membranes of circumscribed cavities, excepting suppuration has taken place ; and, indeed, ulceration in such parts would be a sure forerunner of suppuration.

New formed parts, or such as cannot be said to constitute part of the original animal, as healed sores, calluses of bones, especially those in consequence of compound fractures admit more readily of absorption, especially the progressive, than those parts which were originally formed ; this arises probably from the principle of weakness, and it is from this too, that all adventitious new matter, as tumors, are more readily absorbed than even that which is a substitute for the old. Thus we have tumors more readily absorbed than a callus of a bone, union of a tendon, etc. because they have still less powers than those which are substitutes for parts originally formed.

Ulceration in consequence of death in an external part, takes place soonest on the external edge between the dead and the living. This is visible in the sloughing of parts; for we may observe that sloughs from caustics, bruises, mortifications, etc. always begin at the external edge.

An internal pressure produced by an extraneous body, acts equally on every side of the surrounding parts, and therefore every part being pressed alike, ought, from this cause alone, to produce absorption of the surrounding parts equally on all sides ; supposing the parts themselves similar in structure, or which is the same, equally susceptible of being absorbed ; but we find that one side only of the surrounding living parts is susceptible of this irritation, therefore one side only is absorbed ; and this is always the side which is next to the external surface of the body. We therefore, have always extraneous bodies of every kind, determined to the skin, and on that side to which the extraneous body is nearest, without having any effect, or producing the least destruction of any of the other surrounding parts. From this cause we find abscesses, etc. whose seat is in, or near the centre of a part, readily determined to the surface on the one side, and not on the other ; and whenever the lead is once taken, it immediately goes on. But as some parts, from their structure, are more susceptible of this irritation than others, we find that those parts compofed of fuch ftructure, are often abforbed, although they are not in the fhorteft road to the fkin ; this ftructure is the cellular membrane, as will be taken notice of hereafter.

We find the fame principle in the progrefs of tumors ; for although every part furrounding a tumor is equally preffed, yet the interftitial abforption only takes place on that fide next the external furface, by which means the tumor is, as it were, led to the fkin ; from hence we find that abforption of whole parts more readily takes place, to allow an extraneous fubftance to pafs out of the body, than it will to allow one to pafs in.

Thus we fee, that the flight preffure produced by matter on the infide of an abfcefs has a great effect, and the matter is brought much fafter to the fkin (although very deep) than it would by the fame quantity of preffure applied from without ; and, indeed, fo flight a preffure from without would rather tend to have an oppofite effect, namely, that of thickening.

The reafon of this is evident ; one is, a readinefs in the parts to be freed from a difeafe already exifting ; the other is, a backwardnefs in the parts to admit a difeafe. This principle, therefore, in the animal œconomy produces one of the moft curious phœnomena in the whole procefs of ulceration, viz. the fufceptibility which the parts lying between an extraneous body and the fkin have to ulcerate, while all the other fides of the abfcefs are not irritated to ulceration ; and the neceffity there is that it fhould be fo, muft be very ftriking ; for if ulceration went on equally on all fides of an abfcefs, it muft increafe to a moft enormous fize, and too great a quantity of our folids muft neceffarily be deftroyed.

Bones, we have obferved, are alfo fubject to fimilar circumftances of ulceration ; for whenever an abfcefs forms in the centre of a bone, or an internal exfoliation has taken place, the extraneous body acts upon the internal furface of the cavity, and produces ulceration.

If the matter or dead piece of bone is nearer one fide than the other, ulceration takes place on that fide only ; and here too the provifion of nature in abfceffes comes in, for the adhefive inflammation extends itfelf on the outfide in proportion as ulceration extends itfelf on the infide of the cavity, and as ulceration approaches to the furface of the bone, the adhefive difpofition is given to the periofteum, then to the cellular membrane, etc. And what is very curious, this adhefive inflammation affumes the offifying difpofition, which I have called the offific inflammation, and appears as a fpreading offification, in the fame manner as in the callus of a fimple fracture.

The confequence of thefe two proceffes taking place together in bones is very fingular, for the ulcerative procefs deftroying the infide of a bone while the offifying makes addition to its outfide, the bone often increafes to a prodigious fize, as in cafes of fpinæ ventofæ ; but in the end the ulceration on the infide gets the better, and the matter makes its efcape.

Nature has not only made what might be called an inftinctive provifion in the parts to remove themfelves, fo as to bring extraneous bodies to the fkin for their exit, and thereby, from this principle, has guarded the deeper feated parts ; but has alfo guarded all paffages or outlets, where, from reafoning, we might fuppofe no great mifchief could arife from bringing extraneous bodies thither ; and in many cafes a feeming advantage would be gained ; fuch paffages appearing to be more convenient for the exit of fuch matter, and likely to produce lefs vifible mifchief in procuring them.

Thus a tumor in the cheek, clofe on the internal membrane of the mouth, and fome way from the fkin, fhall in its growth pufh externally, efpecially if there is matter in it, and in time come in contact with the fkin and adhere to it, while it fhall have made no clofer connexion with the fkin of the mouth ; if it fhould fuppurate, and more efpecially if it be of a fcrofulous kind, which is flow in its progrefs, it will break externally ; we even fee abfceffes in the gums opening externally, where the matter has been obliged to go a confiderable way to get to the fkin.

The fame guard is fet over the cavity of the nofe ; if an abfcefs forms in the antrum, frontal finus or faccus lacrimalis, all of which are nearer to the cavity of the nofe than the external furface of the body, ulceration does not follow this fhorteft way, which would be directly into the nofe, but leads the matter to the neareft external furface.

I have feen an abfcefs in the frontal finus, firft attended with great pain in the part, then with inflammations on the whole forehead, at laft matter has been felt under the

ſkin; and on being opened it has led into one or both ſi-
nuſes, and almoſt the whole bone has exfoliated. For ſuch
an abſceſs, the neareſt paſſage would have been directly into the noſe. Abſceſs in the lacrimal ſac, forming what is
called the fiſtula lacrimalis, ariſes alſo from the ſame cauſe;
a curious circumſtance takes place here; but whether peculiar to this part or not, I do not yet know. Beſides
the diſpoſition for ulceration externally at the inner corner
of the eye, there is a defence ſet up upon the inſide, ſo
that the membrane of the noſe thickens very conſiderably;
how far a thickening takes place on the inſide of the noſe,
oppoſite to the antrum, in abſceſſes of that cavity, or
how far it is an univerſal principle in other paſſages, I have
not been able to learn, but am inclined to believe it is not
univerſal. From this principle we can ſee why openings
made into theſe paſſages to make the matter come that way
are more unſucceſsful, than reaſoning (without the knowledge of this principle) would lead us to believe; the opening, therefore ſhould not be made on the inſide (even
where we can do it) excepting the matter is very near, or
elſe the opening ſhould be made very large; and probably,
in ſuch caſes, it may be neceſſary to take out a piece, ſo as
to prevent the uniting proceſs, which is here very ſtrong.

Illuſtrations will be given in other paſſages, when treating of ulceration in general tending to the external ſurface.

III. OF INTERSTITIAL ABSORPTION.

INTERSTITIAL abſorption, I obſerved, was of two
kinds, with reſpect to effect, or rather had two ſtages.
The firſt was, where it took place only in a part, as in the
waſting of a limb in conſequence of its being rendered uſeleſs, whether from diſeaſe in a joint, a broken tendon, or
the dividing of a nerve whereby its influence is cut off; or
where it takes place in the whole body, in conſequence of
ſome diſeaſe, as in acute fever, hectic fever, diabetes, a-
trophy, or the like. The ſecond is the abſorption of a
whole part, where not a veſtige is left. This would ſeem
to be of two kinds; one, where it is only a conſequence of

another difease, and is a neceffary and ufeful effect of that difeafe, as in affifting in bringing parts to the furface; but the other appears to arife from a difeafe in the part it- felf; as the total decay of the alveoli; without any difeafe in the teeth or gums, which in the end fuffer; as alfo a total wafting of a tefticle, the abforption of a callus, etc. It is the firft of thefe two kinds which is moft to my prefent purpofe, and deferves our particular attention. It takes place in a thoufand inftances; we find it gradually taking place in the part of the body which happens to lie between infyfted tumors, and the external furface, when they are making their way to the fkin. This abforption is commonly flow in its progrefs, fo much fo, as even to make the ultimate effect, although con- fiderable, not fenfible till a certain length of time has elapf- ed.

This mode of removing parts, appears to arife from pref- fure, as in the former; but here fome principles are rever- fed. The contents of an incyfted tumor do not give the ftimulus of removal to that fide of the fyft neareft to the external furface, as happens in an abfcefs, fo as to produce a removal of the furface preffed by its contents, which would be the progreffive ulceration, as in our firft divifion; but the tumor gives the ftimulus to the found parts, be- tween it and the fkin, and an abforption of thofe parts takes place, fimilar to that which I fuppofe takes place in the removal of callufes of bones, from weaknefs. We find when- ever an incyfted tumor is formed in the cellular membrane, it in time makes its approaches towards the fkin, by the cel- lular membrane and other parts between it and the fkin being abforbed, fo that the whole fubftance between the cyft and fkin becomes thinner and thinner, till the cyft and the external fkin meet or come in contact, and then in- flammation begins to take place; for as the parts are now foon to be expofed, inflammation takes place to produce a quicker abforption, which borders often upon ulceration. The mode of action in this laft cafe may be, in one refpect, very fimilar to the foregoing folid tumor, for befides the interftitial abforption, the cyft may be looked upon as a tu- mor acting upon, or ftimulating the parts between it and the fkin; therefore the tumor caufes abforption of the con- tiguous cellular membrane, upon which it preffes. This pre- cefs of interftitial abforption of parts is very evident, even in common abfcefs; where a progreffive abforption is going on, it is affifted by this.

I have already obferved, that the interftitial abforption is not attended with, nor produces fuppuration.

IV. OF THE PROGRESSIVE ABSORPTION.

THE firft or principle mode of this action, is the removing of thofe furfaces that are immediatly contiguous to the irritating caufes which is an abforption of neceffity. Thefe caufes, I have obferved, are of three kinds; one, preffure; another, iritating fubftances; and the third, confiderable inflammation on a weak part, efpecially thofe new-formed parts that become a fubftitute for the old. Abforption, from preffure, is the removal of the part preffed, which may arife from a number of caufes. There are tumors, which by preffing upon neighbouring parts, produce it; the preffure of the blood in aneurifms produces it, etc. alfo that furface of an abfcefs which is in contact with the pus, or any other extraneous body; or the ulceration of that part of the furface of the body, which is in contact with a body preffing, as the buttocks or hips of thofe who lie long upon their backs. The heels of many people, who alfo lie long in the fame pofition, as is the cafe with thofe who are under the cure of a fracture of the leg; in which cafe it feems to be a fubftitute for mortification, and is rather a proof of the ftrength of the patient; for if very weak conftitutionally, the fame parts certainly mortify; as alfo the conftant preffure of chains on the legs of prifoners; harnefs on the breafts of horfes.

The fecond of this caufe of abforption is the action of irritating fubftances, fuch as the tears paffing conftantly over the cheeks; as alfo many irritating medicines, producing too much action, and probably at the fame time weakening the parts. The third is, the formation of an ulcer, or fore, on a furface, in confequence of fome difeafe, which has been the caufe of inflammation. Bones are fubject to the fame effect from preffure as the foft parts ; as in confequence of aneurifms ; as alfo from the preffure from tumors: likewife in cafes of the fpina ventofa, where in fome

ᵗhere is nothing to be found in the cavity of the swelling
but blood coagulated ; in others, a grumous or curdly substance.   This blood, or substance, increasing, continues
the pressure, and the inside of the bone is in time absorbed.

I have already observed, that the progressive absorption
is divisible into two kinds ; one without suppuration the o-
ther with.   I shall now observe, that the absorption which
does not produce suppuration, may take place, either from
pressure made by sound parts, upon diseased parts, or
by diseased upon sound parts ; as the effect that the pressure of the coagulated blood has in aneurisms, the moving
blood in the same, which is a sound part, contained in diseased arteries not capable of supporting the pressure of the
moving blood ; as also many tumors, which are diseased
parts, pressing upon natural sound parts, and these diseased parts are simply endowed with life, which I apprehend
makes some difference in the effects respecting the formation
of pus ; also uncommon pressure made by such substances
as are not endowed with any irritating quality, sufficient
to produce the suppurative inflammation, as a piece of
glass, a lead bullet etc. all of which I shall now more fully
explain.

Of this first division, viz. from pressure without suppuration, we have several instances ; in aneurisms; especially
when they are in the aorta, and principally at the curve ;
and when arrived at a considerable size, so as to press a-
gainst the surrounding parts, particularly against the back-
bone, as also against the sternum ; all of which will be according to the situation of the aneurism ; we find in such
cases, that from the dilatation of the artery (which arises
from the force of the heart) the artery is pressed against those
bones, and that the substance of the artery in the part
pressed is taken into the constitution.   This absorption begins at the external surface of the artery, where it comes
in contact with the bone, and continues there till the whole
artery is absorbed ; then the bone itself comes in contact
with the circulating blood, and not being naturally intended to be washed by moving blood, the bone or bones are also absorbed from this pressure, and motion of the blood a-
gainst them.   The adhesive or strengthening disposition
takes place in the surrounding parts, and is of great service
here, as it unites the circumference of the unabsorbed part
of the artery to the surrounding parts ; as also the cellular

membrane beyond the furface of abforption, (when in foft parts) fimilar to the preceding adhefive inflammation go- ing before ulceration in an abfcefs; but it is here much ftronger, for ftrength is wanted as well as adhefion while it is dilating; fo that a cavity of fome ftrength for the mov- ing blood is always kept entire, and no extravafation can take place, nor can the parts readily give way.

Another inftance of this abforption occurs in thofe cafes where living tumors make their way to the fkin without the formation of an abfcefs. I once faw a remarkable inftance of this in a Highland foldier, in the Dutch fervice, who had a folid tumor formed, either in the fubftance of the brain, or, what is more probable, upon it, viz. in the pia mater, for it feemed to be covered by that membrane : the tumor was oblong, above an inch thick, and two or more inches long; it was funk near its whole length into the brain, feemingly by the fimple effects of preffure; but the outer end of it, by preffing againft the dura mater, had pro- duced the abforbing difpofition in that membrane, fo that this membrane was entirely gone at that part.

The fame irritation from preffure had been given to the fcull, which alfo was abforbed at this part; after which, the fame difpofition was continued on to the fcalp.

As thefe refpective parts gave way, the tumor was pufh- ed further and further out, fo that its outer end came to be in this new paffage the abforbents were making for it in the fcalp, by which it probably would have been difcharged in time, if the man had lived; but it was fo connected with the vital parts, that the man died before the parts could re- lieve themfelves; while all thefe exterior parts were in a ftate of abforption, the internal parts which preffed upon the inner end of the tumor, and which preffure was fufficient to pufh it out, did not in the leaft ulcerate, nor did the tu- mor itfelf, which was preffed upon all fides, in the leaft give way in its fubftance. No matter was to be obferved here from either the dura mater, the unconnected edge of the bones of the fcull, nor from that part of the fcalp which had given way; and, perhaps, the reafon was, the tumor being a living part, and not an extraneous one. The general effect was, however, fimilar to the progrefs of an abfcefs, infomuch that it was on that fide neareft to the external furface of the body that the irritation for abforption took place.

The firft fpecies of the abforption of whole parts is feldom or never attended with pain.   Its progrefs is fo very flow, as to keep pace with our fenfations, and in many cafes it is not even attended with inflammation.

I believe that this abforption feldom or ever affects the conftitution, although, in fome cafes it takes its rife from affections of the conftitution, as in the cafes of the abforption of callus.

V. OF ABSORPTION ATTENDED WITH SUPPURATION, WHICH I HAVE CALLED ULCERATION.

I SHALL now give an account of that part of the actions of the abforbing fyftem, which I call ulceration, and which is the fecond of our firft divifion, refpecting the formation of pus, viz. that which is connected with the formation of that fluid, being either a confequence of it, or producing it, and is that which in all cafes conftitutes an ulcer.   It is this which principally conftitutes the progreffive abforption *

This differs from the foregoing in fome circumftances of its operations.   It either takes place in confequence of fuppuration already begun, and then the pus acts as an extraneous body, capable of producing preffure ; or abforption attacks external furfaces from particular irritations or weaknefs, in which cafe fuppuration, forming an ulcer, muft

* I have given it the term ulceration, becaufe ulcer is a word in ufe to exprefs a fore, and it is by this procefs that many ulcers are formed.   The operations produced in ulceration, have not hitherto been in the leaft underftood, therefore a very erroneous caufe of thefe operations has been always fuppofed.   It has always been fuppofed that thofe folids which were vifibly gone, were diffolved into pus : from whence arofe the idea of matter being compofed of folids and fluids, which we have endeavoured to refute.

follow, let the caufe of that breach or lofs of fubftance be
what it may.

In order to produce ulceration from preffure, I may again
take notice, that it requires a much greater preffure from
without than from within ; and when it is from within,
the ulceration is quicker, when near to the fkin, than when
deep or far from it ; the nearer to the fkin, the inflamma-
mation more readily takes place ; and I have alfo obferved,
that inflammation, although it takes place in deep feated
parts, yet it feldom or never extends deeper, but approaches towards the external furface ; and as inflammation feems
to proceed and is effential to this procefs, we fee the reafon why it fhould take place fooner if near to the fkin, and
go on fafter the nearer it comes to it.

The procefs of ulceration which brings matter to the external furface is not wholly the abforption of the inner furface of the abfcefs, for there is an interior or interftitial abforption of the parts lying between the inner furface of
abfcefs and the fkin, fimilar to the approach of incyfted tumors, as has been defcribed. And befides this affiftance,
I have already obferved, there is a relaxing and elongating procefs carried on between the abfcefs and the fkin ;
and at thofe parts only where the matter appears to point.

This procefs of ulceration, or abforption, with fuppuration, is almoft conftantly attended by inflammation ; but
it cannot be called an original inflammation, but a confequent, which gave rife to the term ulcerative inflammation.
It is always preceded by the adhefive inflammation, and
perhaps it is fimply this inflammation which attends
it ; we find the adhefions produced anfwering very
wife purpofes ; for although the adhefive inflammation
has preceded the fuppurative, and of courfe all the parts
furrounding the abfcefs are united, yet, if this union of the
parts has not extended to the fkin, where the abfcefs or
matter is to be difcharged, in fuch a cafe, wherever the ulceration has proceeded beyond the adhefions there the matter will come into unadhering parts ; the confequence of
which will be, that the fluid, or matter, will diffufe itfelf into the cellular membrane of the part, and from thence over
the whole body, as in the eryfipelatous fuppuration ; but
to prevent this effect, the adhefive inflammation takes the
lead of ulceration. There are many other caufes of ulceration, which take place on furfaces, where we do not fee

the fame neceffity for it ; when the matter formed can be, and is difcharged without it ; fuch parts are many old fores ; the infide of the ftomach and inteftines, and indeed all the furfaces abovementioned, which do not admit readily of the the adhefive inflammation ; under fome circumftances admit of the ulcerative. This effect would appear to arife from the violence of the inflammation, the parts being fo weakened, either by it, or fome former difeafe, that they can hardly fupport themfelves ; for we find in falivations, where the whole force of the mercury has been determined to the mouth, they have become weakened by long and violent action ; the gums and infide of the mouth will ulcerate ; alfo, from the fame weakening difpofition, the gums will ulcerate in bad fcurvies ; therefore weaknefs joined with inflammation, or violence of action, appears to be the immediate caufe in fuch cafes.

The effect then of irritation, as above defcribed, is to produce firft the adhefive inflammation in fuch parts as will readily admit of it, and if that has not the intended effect, the fupurative takes place, and then the ulceration comes on to lead the matter already formed to the fkin, if it is confined.

The natural confequence of fuppuration in fuch parts, is the growth of new flefh, called granulations, which are to repair the lofs the parts fuftained by the injury done ; but in all outlets, where the adhefive would be hurtful, the irritation firft only produces the fuppurative inflammation; but if carried further, the adhefive will take place, as has been defcribed ; and, as in fuch parts the matter formed has an outlet, ulceration is alfo avoided ; and, as in fuch cafes no parts are deftroyed, granulations are alfo excluded.

There appears to be a curious circumftance attending ulceration, which is the readinefs with which it feems to abforb every other fubftance applied to it, as well as the body itfelf ; at leaft this appears to be the cafe with the fmallpox after inoculation ; as alfo the venereal chancre; whether arifing from the abforbents at the time being in the act of abforbing, or whether they promifcuoufly abforb what is applied, along with the part themfelves. In fuch cafe it might be a queftion alfo, whether the parts of the body which they do abforb have the fame difpofition with the pus of that part, as in the cancer, therefore contaminate the conftitution, as in the fmall-pox and venereal difeafe, as readily as if it was the pus.

From what has been obferved, it muft appear that any irritation which is fo great as to deftroy fuddenly the natural operations of any one part, and the effect of which is fo long continued as to oblige the parts to act for their own relief, produces in fome parts, firft the adhefive inflammation ; and if the caufe be increafed, or continue ftill longer, the fuppurative ftate takes place, and all the other confequences, as ulceration ; or, if in the other parts, as fecreting furfaces, then the fuppurative takes place immediately ; and if too violent, the adhefive will fucceed ; or if parts are very much weakened, the ulcerative will immediately fucceed the adhefive, and then fuppuration will be the confequence.

This fpecies of ulceration in general gives confiderable pain, which pain is commonly diftinguifhed by the name of forenefs ; this is the fenfation arifing from cutting with an inftrument, which operation is very fimilar to ulceration ; but this pain does not attend all ulcerations, for there are fome of a fpecific kind, which give little or no pain, fuch as the fcrofula ; but even in this difeafe, when the ulceration proceeds pretty faft, it gives often confiderable pain ; therefore the pain may be in fome degree proportioned to the quicknefs of its operation.

The greateft pain which in general attends this operation arifes from thofe ulcerations which are formed for the purpofe of bringing the matter of an abfcefs to the fkin : as alfo where ulceration begins upon a furface, or is increafing a fore. Whether the increafe of pain arifes from the ulcerative inflammation, fingly, or from the adhefive and ulcerative going on together in the fame point, is not eafily determined ; but in fome cafes thefe three are pretty rapid in their progrefs, and it is more than probable that the pain arifes from all thefe caufes.

In thofe cafes where ulceration is employed in feparating a dead part, fuch as floughing, exfoliation, etc. it is feldom attended with pain ; perhaps it may not be eafy to affign a caufe for this.

The effects that ulceration has upon the conftitution I have mentioned, with the effects that other local complaints have upon it.

It is eafy to diftinguifh between a fore that is ulcerating, and one which is ftanding ftill, or granulating.

The ulcerating fore is made up of little cavities or hollows, and the edge of the fkin is fcolloped or notched ; is thin, turned a little out, and overhangs, more or lefs, the fore. The fore is always foul, being probably compofed of parts not completely abforbed ; and difcharges a thin matter.

But when the ulceration ftops, the edge of the fkin becomes regular, fmooth, a little round or turned in, and of a purple colour, covered with a femi-tranfparent white.

VI. OF THE RELAXING PROCESS.

Besides thefe two modes of removing whole parts, acting fingly or together, there is an operation totally diftinct from either, and this is a relaxing and elongating procefs, carried on between the abfcefs and the fkin, and at thofe parts only where the matter appears to point. It is poffible that this relaxing, elongating, or weakening procefs may arife in fome degree from the abforption of the interior parts ; but there is certainly fomething more, for the fkin that covers an abfcefs is always loofer than a part that gives way from mere mechanical diftention, excepting the increafe of the abfcefs is very rapid.

That parts relax, or elongate, without mechanical force, but from particular ftimuli, is evident in the female parts of generation, juft before the birth of the foetus ; they become relaxed prior to any preffure. The old women in the country can tell when a hen is going to lay, from the parts becoming loofe about the anus.

That this relaxing procefs takes place between an abcefs and the fkin is evident in all cafes, but was more demonftratively fo in the following cafe than commonly can be obferved where an increafe of furface takes place without the vifible lofs of fubftance, for here both could be exactly afcertained ; and, indeed, no abfcefs could fwell outwards, excepting by diftention, without it.

In the following cafe this procefs was particularly evident.

A lad about thirteen years of age, was attacked with violent inflammation in his belly, without any apparent caufe. The ufual means were ufed, but without effect. His belly began to fwell in a few days after the attack, and his fkin became cold and clammy, efpecially his feet and hands. Once, when he made water, it was tranfparent like fpring water, with a little cloud of mucus. In feveral places of the belly, there appeard a pointing, as if from matter; one of thofe, which was juft below the fternum, became pretty large, and difcoloured with a red tint. Although there was not any undulation or perfect fluctuation (there not being fluid enough for fuch a feel) yet it was plain there was a fluid, and moft probably from the pointings, it was matter in confequence of inflammation, and that it was producing ulceration on the infide of the abdomen for its exit; therefore it was thought advifable, as early as poffible, to open the belly at one of thofe parts. I made a fmall opening into the pointing part, juft below the fternum, hardly an inch long: when I was performing the operation, I faw plainly the head of the rectus mufcle, which I cut through in the direction of its fibres. There was immediately difcharged by this wound about two or three quarts of a thin bloody matter. The fwelling of the abdomen fubfided of courfe; his pulfe began to rife and become more full and foft; and his extremities became warmer; he was ordered bark, etc. but he lived only about fixty hours after the operation.

On opening his abdomen after death we found little or no matter lying loofe; all had made its efcape through the wound. The whole inteftines, ftomach, and liver, were united by a very thick covering of the coagulating lymph, which alfo paffed into all the interftices between them, by which means they were all united into one mafs, the liver alfo adhered to the diaphragm, but none of the vifcera adhered to the infide of the belly on its fore part, for there the matter had given the ftimulus for ulceration, which prevents all adhefions. The procefs of ulceration had gone on fo far as to have deftroyed the whole of the peritoneum on the fore part of the abdomen, and the tranfverfales, and recti mufcles, were cleanly diffected on their infide.

The tendons of the lateral mufcles that pafs behind the heads of the recti, were in rags, partly gone, and partly in the form of a flough.

From this view of the cafe, we muft fee how nature had guarded all the moft effential parts. In the time of the adhefive ftage, fhe had covered all the inteftines with a coat of coagulating lymph, fo as to guard them; and this, probably, upon two principles, one, from their being canals; and therefore loath to admit of penetration in that way; the other, from their being more internal than the parietes of the abdomen; one fide is therefore thickened for their defence, while the other is thinned for the relief of the part.

Here the cavity of the abdomen had affumed all the properties of an abfcefs, but it was fo connected with the vital parts, which alfo fuffered much in the inflammation, that the patient could not fupport the neceffary proceffes towards what would be called a radical cure in many other parts; and indeed, confidering the mifchief done to the abdomen and its vifcera, it is aftonifhing he lived fo long.

The moft curious circumftance that happened, was the appearance of pointing in feveral places; for why one part of the abdomen fhould have pointed more than another is not eafily accounted for, fince every part of the anterior portion was nearly equally thin, each part was equally involved in the abfcefs, and the ulceration had not yet begun with any of the mufcles. To account for this, let us fuppofe that one, two, or three parts (by fome accident) were more fufceptible of the ulcerative ftimulus than the others, and that the parts were ready to give way; but although thefe parts which were pointing, were the places where ulceration would have gone on brifker, yet it had not proceeded further here than in any other part; it had only gone through the peritoneum, and the tendons of the broad mufcles; and the recti mufcles were found and perfect at the place where I made the opening, which was the moft protuberant of any; therefore this pointing did not appear to arife from weaknefs or thinnefs of this part; and, even fuppofing that the pointing was an effect of weaknefs, it would imply a great deal of preffure on the infide, (which at leaft was not the cafe here) and fimple preffure, although a hundred times greater, which we often fee take place in dropfies, would not produce a pointing, if not attended with fome fpecific power.

If preffure then was not fufficient to produce this effect in the prefent cafe, and if the parts which pointed were as mechanically firm as at any other, to what other caufe can we attribute the diftention of this part, but to the weakening, elongating, and relaxing procefs, which I have already defcribed.

This obfervation of the relaxing procefs going on in the fubftance of the parts where it points, is verified in a thoufand inftances : fuppofe a large abfcefs in the thigh, only covered by the fkin and adipofe membrane, which fhall go on for months without producing ulceration, and of courfe not point any where, but fhall be a fmooth, even, and uniform furface, let it receive the ftimulus of ulceration in any one part, that part will immediately begin to point, although it may be thicker there than at fome other parts of the fame abfcefs.

The preffure neceffary to allow extraneous matter to make its efcape, need not be great ; for in many abfcefs which have been opened, or have opened of themfelves, but not at the moft depending part, fo that the matter is allowed to ftagnate at the lower part of the cavity, making a very flight preffure, we find that this alone is fufficient to produce ulceration in that part, and of courfe a frefh opening is produced, more efpecially if near the fkin ; this we fee often takes place in abfceffes of milk breafts, when the opening is not at a depending part, and appears to be common in the fiftula in ano ; for it frequently happens that the ulceration goes on at firft towards the gut ; but before this has taken place, ulceration has gone on fome way by the fide of it, to bring the matter externally, which weight of matter is alone fufficient to continue the fame procefs.

Vol. II.   A a

VII. OF THE INTENTION OF ABSORPTION OF THE BODY IN DISEASE.

THIS, like every thing elfe in nature, involves in it two confequences, the one beneficial, the other hurtful ; both of which this has in a confiderable degree : however, if we underftood thoroughly all the remote caufes we fhould probably fee its utility in every cafe, and thefe effects, however bad in appearance, yet are neceffary, of courfe in the end falutary. The ufe arifing from what may be called the natural abforption of parts, fuch as the forming or modelling procefs, as alfo the abforption of parts become unfit for the new mode of life, as the abforption of the thymus gland, etc. is involved in its neceffity, and belongs to the natural hiftory of the animal; but that arifing from difeafe is directly to the prefent purpofe. In the hiftory I have juft given, its ufe muft, I apprehend, be evident; for we plainly fee, in each mode of abforption, it often produces very falutary effects ; and we may fay, that although it often arifes from difeafe, yet its operations and effects are often not at all a difeafe ; and, probably, in thofe cafes where we cannot affign a caufe, as in wafting of parts, atrophy, etc. yet it is moft probable that its ufe is confiderable. It is likely, that under fuch a difeafe, or ftate of body, or parts, it would be hurtful to have them full and ftrong ; where it produces a total wafte of a part, its utility is probably not fo evident ; but in the progreffive abforption, where it is leading bodies externally, or in confequence of fuppuration, where it is bringing matter externally, its ufe is plain ; or even in the formation of an ulcer, or the fpreading of an ulcer, its ufe may be confiderable. I have formerly called it the natural furgeon ; and where it can do its bufinefs it is in moft cafes preferable to art : this is fo evident in many cafes, that it has been a conftant practice to attempt to promote it, in bringing abfceffes to the furface, and in the exfoliation of bone, etc. although not accounted for upon the principle of abforption, yet the effect was vifible, and its ufe allowed.

VIII. THE MODES OF PROMOTING ABSORP-
TION.

THE hiftory that was given of the caufes of abforption, in fome degree explains the modes of promoting it ; but as there were fome natural caufes which we cannot imitate, it is principally thofe that can be rendered ufeful that we are to take notice of in this place.

To promote abforption of the body itfelf, is no difficult operation ; it is only to leffen the fupply, and increafe the wafte, which laft is often done by medicine ; or to take fuch things as will render the fupply lefs efficacious, as vine-gar or foap ; but probably thefe act principally on the fat: to promote abforption of difeafed parts, or parts increaf-ed, or parts newly formed, is not fo eafy a tafk, although the latter may be the moft eafy of the whole; for I have af-ferted, that newly formed parts are weaker in their living powers than the original formed parts ; this, in fome de-gree, gives us a hint; for if we have a mode of producing a wafte of the orignal body, under this general wafte, new formed parts muft fuffer in a degree proportional to their weaknefs ; and therefore will fuffer a diminution in the fame proportion; but this is too often not fufficient, or at leaft what would be fufficient for the difeafe would be too great for the conftitution to bear : however, we find in particu-lar cafes, that this practice has fome effect ; probably the beft debilitating medicine is mercury, and it probably may act in more ways than one. It may promote abforp-tion from a peculiar ftimulus, producing neceffity, or a ftate under which the parts cannot exift. Electricity, and moft other ftimulents, probably act in the fame manner ; for we find that violent inflammation is often a caufe. Death in a part is fure to promote abforption, in order to produce a feparation of the dead parts : and we even find that a part being difeafed gives a tendency to feparation, and only re-quires a confiderable inflammation to promote it, fuch as warts coming away in confequence of inflammation. A difeafed part has fuch power of giving the proper ftimulus to the adjacent found part, that if injured, or rendered dead in part by the application of a cauftic, for inftance, the found part underneath will begin to relax, and fhew more diftinctly the limits or boundaries of the difeafe ; fo that a feparation of the difeafed parts begins to take place, although the cauftic has not reached nearly fo far, and may give us an intimation of the extent of the difeafe, which we could not get before. It is in fome meafure upon this principle that arfenic removes tumors which extends beyond the immediate effect of the medicine.

Preffure is one of the caufes of abforption in general, particularly the progreffive, which, in the refolution of parts, is not the mode wanted ; but it alfo affifts in producing the interftitial ; and if it could be made to produce the fecond of the interftitial, viz. abforption of the whole, as in the total decay of the thymus gland, then it would be fufficient in thofe cafes where it could be applied : but the preffure muft be be applied with great care ; for too much will either thicken or ulcerate, which laft may be a mode of abforption we do not want : however, thefe effects will happen according to circumftances ; for I have an idea, that entirely new formed parts, as tumors, will not be made to thicken by preffure, therefore may be preffed with all the force the natural furrounding parts will allow. On the other hand there are many cafes where we would wifh to prevent abforption ; but when this is the cafe, we fhould be certain that the part which was to have been abforbed is fuch as can be rendered ufeful afterwards ; of which I have my doubts in many cafes.

IX. ILLUSTRATIONS OF ULCERATION.

Now that I have been endeavouring to give ideas of thefe effects of inflammation, viz. adhefion, fuppuration, and ulceration, let me next mention fome cafes which frequently occur, as illuftrations, which will give a perfect idea of thefe three inflammations : and, for the clearer underftanding them, I fhall illuftrate them upon the inflammation, fuppuration, and ulceration of the large circumfcribed cavities. For inftance, an inflammation attacks the external coat of an inteftine ; the firft ftage of this inflammation produces adhefions between it and the peritoneum lining the abdominal mufcles : if the inflammation docs not ftop at this ftage, an abfcefs is formed in the middle of thefe adhefions, and the matter acts as an extraneous body ; the abfcefs increafing in fize from the accumulation of matter, a mechanical preffure is kept up, which irritates, and the fide next to the fkin is only fufceptible of the irritation ; this irritation not deftroying the difpofition to form matter, fuppuration is ftill continued, and the ulcerative inflammation takes place.

If fuppuration began in more parts of the adhefions than one, they are commonly united into one abfcefs ; an abforption of the parts between the abfcefs and the fkin takes place, and the matter is led on to the external furface of the body, where it is at laft difcharged.

If the difpofition for ulceration was equal on every fide of the abfcefs, it muft open into the inteftine, which is feldom the cafe, although it fometimes does ; for the fame precautions are not taken here as in many other fituations; for in fome others, as in the nofe, in the cafe of an abfcefs of the lacrimal fack, the paffage is thickened towards the nofe. In the cafe above-defcribed, however, the abdominal mufcles, fat, and fkin are removed, rather than the coats of the inteftine. Cafes of this kind have come under my own obfervation.

In this cafe, if adhefions had not preceded ulceration, the matter muft have been diffufed over the whole cavity of the belly ; if the adhefive inflammation had not likewife gone before the ulceration in the abdominal mufcles, etc. the matter would have found a free paffage from the abfcefs into the cellular membrane of the abdomen, as foon as the ulceration had got through the firft adhefions, as is often the cafe in eryfipelatous fuppurations.

Abfceffes between the lungs and the pleura, in the liver, gall-bladder, etc. rife to the furface from the fame caufe ; alfo in lumbar abfceffes, where one would at firft imagine the readieft place of opening would be the cavity of the abdomen, or gut; the parts neareft to the fkin are removed, and the matter paffes out that way ; however, in abfceffes fo very deep, it does not always happen that one

fide only is fufceptible of the irritation, and we fhall find
that the matter is taking different courfes.

Abfceffes in the fubftance of the lungs fometimes differ
from the above defcribed ; for they fometimes open into
the air-cells : it is, becaufe the adhefive inflammation finds
it difficult to unite the air-cells, and branches of the trachea, (as was defcribed in treating of that inflammation)
and alfo in the fubftance of the lungs, it may be difficult
to fay where it can take a lead externally, from which,
probably, the air-cells become fimilar to an external furface,
and then ulceration takes place on that fide of the abfcefs
which is neareft to the cells ; therefore we find that the matter
gets very readily into the air-cells, and from thence into
the trachea.

That the air-cells do not take on the adhefive ftate is e-
vident in moft abfceffes in this part; for we find in moft of
thofe cafes that the air cells are expofed, as alfo the branches of the trachea, and the parts of the lungs which com-
pofe this abfcefs have not the firmnefs and folidity which
the adhefive inflammation generally produces in thofe parts
where it takes place.

Thus too we find it going on in large abfceffes, even after
they have been opened, but are fo fituated or circumftan-
ced as to have fome part of the abfcefs on that fide immediately under the fkin preffed by fome other part of the body which lies underneath. For inftance, when a large abfcefs forms on the outer, and upper part of the thigh, oppofite the great trochanter, which is a very common complaint, and an opening is made into it, or it burfts below,
or on the fide of that bone, but not directly oppofite to the
trochanter itfelf, in fuch cafes it frequently happens, that
the preffure of the trochanter on the infide of the abfcefs,
viz. the cellular and adipofe membrane, and the fkin covering the trochanter, that this preffure produces ulceration of
thefe parts ; which procefs is continued on through the
fkin, and makes a fecond opening directly upon the trochanter.

It is curious to remark, how thefe proceffes of nature
fulfil their appointed purpofes, and go on no further ; for
any young flefh, or granulations, which may have formed
upon the trochanter, which very often happens before this
ulceration is completed, yet thefe do not ulcerate, although

the preſſure was as great, or greater upon them than it was upon the parts which gave way.

This is upon the principle, that preſſure from without has not the ſame effect as from within. The fiſtula lau-rimelis is another ſtrong proof of ulceration only taking place towards the external ſurface, and ſecuring the deep-er ſeated parts ; as alſo the ulceration in conſequence of matter in the frontal ſinuſes.

An effect of the ſame kind we have obſerved in milk-breaſts. In theſe caſes the ſuppuration commonly begins in many diſtinct portions of the inflamed parts, ſo that it is not one large circumſcribed abſceſs, but many ſeparate ſi-nuſes are formed, all of which generally communicate : now it uſually happens, that only one of theſe points exter-nally ; which being either opened or allowed to break, the whole of the matter is to be diſcharged this way ; but it frequently happens, that the matter does not find a ready outlet by this opening, and then one or more of theſe dif-ferent ſinuſes make diſtinct openings for themſelves ; which ſhews how very eaſily the ſlight preſſure of ſuch a trifling confinement of matter can produce the ulcerative inflam-mation. Ulceration is therefore no more than an opera-tion of nature to remove parts out of the way of all ſuch preſſure as the parts cannot ſupport ; and accordingly it be-gins where the greateſt preſſure is felt, joined with the na-ture of the parts and its vicinity to the ſkin.

It is curious to obſerve that the ulcerative proceſs has no power over the cuticle, ſo that when the matter has got to that part it ſtops, and cannot make its way through, till the cuticle burſts by diſtention ; but in general the cuticle is ſo thin as to give but very little trouble * : however,

---

* This is the reaſon why many abſceſſes in the palms of the hands, ſoles of the feet, fore part of the fingers, and about the nails, commonly called whitloes, etc. more eſpecially in work-ing people give ſo much pain in the time of inflammation, and are ſo long in breaking, even after the matter has got through the cutis to the cuticle ; the thickneſs of the cuticle, as alſo the rigidity of the nail, acting in thoſe caſes like a tight bandage, which does not allow them to ſwell or give way to the extravaſation ; for in the cuticle there is not the relaxing power, which adds conſiderably to the pain ariſing from the inflammation ; but when the abſceſs has reached to this thick cuticle it has no power of irritation, and therefore acts only

in many places it is fo thick as to be the caufe of very trou-
blefome confequences.

Thus far I have confidered ulceration as arifing from vifible irritations, joined with a fufceptibility of the parts
for fuch particular irritation ; but, befides thofe above-defcribed, we often have inftances of ulceration taking place
from a difpofition in a part, and where perhaps no reafon
can be affigned, but weaknefs in the part.   I obferved before, that fome parts of the body were more fufceptible of
ulceration than others.   I then fpoke of original parts ;
but I now remark that, newly formed parts are much
more fufceptible of ulceration than the original ;   fuch as
cicatrices, granulations, callufess, etc. for we find this
difpofition often taking place in old cicatrices from very flight

by diftention ; and this is in moft cafes fo confiderable, as to
produce a feparation of the cuticle from the cutis, for a confiderable way round the abfcefs ; for I obferved when on
inflammation, that it commonly produced a feparation of the
cuticle ; all of which circumftances taken together make thefe
complaints much more painful than a fimilar fized abfcefs
in any other foft part.   The application of poultices, in thefe
cafes, is of more benefit than in any other, becaufe here they
can act mechanically, viz. the moifture being imbibed by the
cuticle, as in a fponge, and thereby foftening the cuticle, by
which means it becomes larger in its dimenfions and lefs dura-
ble in its texture.   Thefe cafes fhould be opened as foon as poffible, to avoid the pain arifing from diftention, and the feparation of the cuticle ; when it is conceived it means to point at any
one part, paring off the thick cuticle, near the cutis, is allowing the matter to make its efcape more readily, when it has not
through the cutis.   There is a circumftance which almoft always attends the opening fuch an abfcefs, viz. the foft parts underneath pufh out through the opening in the cuticle, like a tur-
gus, which, when irritated from any accident, give a greater
idea of forenefs perhaps than any other morbid part of the machine ever does : this is owing to the furrounding belts of cuticle not having given way to the increafe of the parts underneath,
by which means they are fqueezed out of this fmall opening,
like paint out of a bladder.   It is a common practice to
eat this down by efcharotics, as if it was a difeafed fungus ;
but this additional pain is very unneceffary, as the deftroying a
part which has only efcaped from preffure, cannot in the leaft
affect that which is within ; and by fimply polticing till the
inflammation, and of courfe the tumefaction fubfides, thefe protruded parts are gradually drawn into their original fituations.

caufes ; fuch as irregularity in the way of life, or violent exercife, which is feen every day in our hofpitals, where the parts feem incapable of fupporting themfelves. Remarkable inftances of this are recorded in Anfon's Voyages, where the habit was fo much debilitated, as to allow all the old fores to ulcerate, or break out anew: the callufes were abforbed and taken into the circulation; and we alfo find, that, all thefe parts perform the operation of floughing when dead, much fooner than original parts.

Now it is evident, in thefe cafes mentioned in Anfon's Voyages, that the whole frame of body was weakened by the hardfhips fuffered in this expedition ; and that the young, or new formed fubftances would fuffer in a greater degree, arifing from their being lefs firm and fixed than that which had been an original formation, and fubfifted from the firft ; and, as no repaired parts are endowed with the powers of action or refiftance, equal to an original part, it is no wonder that this new flefh, fharing in the general debility, became incapable of fupporting its texture : perhaps a very fenfe of this debility proved an irritation, or the caufe of that irritation which produced the abforption of parts ; however, that may be, it is a general fact that, parts which are not originally formed, commonly give way fooner in depravations of the habit : in like circumftances, alfo, old fores that are healing, will break out, fpread, and undo, in twenty-four hours, as much of the parts as has been healing in fo many weeks.

All thefe obfervations tend to prove, that new formed parts are not able to refift the power of many difeafes, and to fupport themfelves under fo many fhocks, as parts originally formed ; which will be ftill further illuftrated, in treating of the power of abforption.

I obferved that, although a part is lofing ground or ulcerating, yet it continues fupparating ; for while a matter-forming furface is ulcerating, (whether an original formed part of the body, fuch as in moft abfceffes, or a new formed fubftance, fuch as granulations) we find that it ftill fecretes pus.

In fuch cafes the adhefive inflammation proceeds very rapidly, and would feem to prepare the parts as it goes for immediate fuppuration the moment they are expofed.

VOL. II. B b

CHAPTER VII.

GRANULATIONS.

W E come now to trace the operations of nature in bringing parts whofe difpofition, action, and ftructure, had been preternaturally altered, either by accident, or difeafed difpofitions, as nearly as poffible to their original ftate. In doing this we are to confider the conftitution, and the parts as free from difeafe ; becaufe all actions which tend to the reftoration of parts are falutary ; the animal powers being entirely employed in repairing the lofs, and the injury, fuftained both from the caufe, and arifing from the courfe of the immediate effects, viz. inflammation, fuppuration, and ulceration : now fuch operations cannot certainly be looked upon as morbid.

Nature having carried thefe operations for reparation fo far, as the formation of pus, fhe, in fuch cafes, endeavours immediately to fet about the next order of actions, which is the formation of new matter, upon fuch fuppurating furfaces as naturally admit of it, viz where there has been a breach of folids, fo that we find, following, and going hand in hand with fuppuration, the formation of new folids, which conftitute the common furfaces of a fore. This procefs is called granulating, or incarnation ; and the fubftance formed, is called granulation.

Granulations have I believe, been generally fuppofed to be a confequence of, or always an attendant on fuppuration : but the formation of granulations is not confined to a breach of folids where the parts have been allowed to fuppurate, as either from accident, or a breach of the folids in confequence of an abfcefs ; but it takes place under other circumftances ; for inftance, when the firft and fe-

cond bond of union has failed, as in fimple fractures which
will be noticed hereafter.

Suppuration, I obferved, arofe in confequence of an injury
having been done to the folids, fo as to prevent them, for
fome time, from carrying on their natural functions; and
I alfo obferved that, it was immaterial whether this injury
had expofed their furfaces, as in cafes of accidents
and wounds; or whether the furfaces were not expofed, as
in cafes of abfceffes in general; for in either of them fuppuration would equally take place; I likewife obferved that,
it was not neceffary that there fhould be a breach in the
continuity of parts for fuppuration to take place in many
cafes, becaufe all fecreting furfaces were capable of fuppuration; but this laft feems not to be fo commonly the
cafe with granulations. I believe that no internal canal
will granulate, in confequence of fuppuration, except there
has been a breach of furface, and then it is not the natural urface which granulates, but the cellular membrane,
etc. as in other parts.

Wounds that are kept expofed do not granulate till inflammation is over, and fuppuration has fully taken place;
for as the fuppurative inflammation conftantly follows
when wounds come to be under fuch circumftances, it
would feem to be in fuch cafes a leading and neceffary procefs for difpofing the veffels to granulation.

Setting out then with the fuppofition, that this inflammation is in general neceffary, under the above circumftances,
for difpofing the veffels to form granulations we fhall at
once fee how it may operate in the fame manner whether
it arifes fpontaneoufly from a wound, the laceration of parts,
mortification, bruife, c. or in fhort any other power
which deftroys or exp es the innumerable internal cells, or
furfaces, fo as to prevent their carrying on their natural
functions.

Few furfaces, in confequence of abfceffes, granulate till
they are expofed; fo that few or no abfceffes granulate till
they are opened, either of themfelves or by art; and therefore in an abfcefs, even of very long ftanding, we feldom
or ever find granulations. In abfceffes, after they have
been opened, there is generally one furface that is more
difpofed to granulate than the others, which is the furface
next to the centre of the body, in which the fuppuration

took place. The furface next to the fkin hardly ever has the difpofition to granulate : indeed, before opening, its action was that of ulceration, the very reverfe of the other: but even, after opening, that fide under the fkin hardly granulates, or at leaft not readily. I may farther obferve that expofure, is fo neceffary to granulation, even on fuch furfaces as arife from a broken continuity of parts, that if the abfcefs is very deep feated, they will not granulate kindly, without being freely expofed, which alone often becomes a caufe why deep feated abfceffes do not heal fo readily, and often become fiftulous.

Upon the fame principle of granulations forming more readily upon that furface which is next to the centre, or oppofite to the furface of the body, is to be confidered their tendency to the fkin. Granulations always tend to the fkin, which is exactly fimilar to vegetation ; for plants always grow from the centre of the earth towards the furface ; and this principle was taken notice of when we were treating of abfceffes coming towards the fkin.

I. OF GRANULATIONS, INDEPENDENT OF
SUPPURATION.

The formation of granulations, I have obferved, is not wholly confined to a breach made in the folids, either by external violence and expofure, or in confequence of a breach in the folids, which had been produced by fuppuration and ulceration, and afterwards expofed ; for parts are capable of forming granulations, or what I fuppofe to be the fame thing, new animal matter, where a breach has been made internally, and where it ought to have healed by the firft intention ; but the parts being baulked in that operation, often do not reach fo far as fuppuration, fo as to produce the moft common caufe of granulation. The firft inftance of the kind that gave me this idea, was in a man who died in St. George's Hofpital.

January 1777.   A man about fifty years of age, fell and broke his thigh-bone, nearly acrofs, and about fix inches above the lower end.   He was taken into St. George's Hofpital; the thigh was bound up, and put into fplints, etc. The union between the two bones did not feem to take place in the ufual time.   He was taken ill with a complaint in his cheft, which he had been fubject to before, and died between three and four weeks after the accident.

On examining the parts after death, there were found little or no effects of inflammation in the foft parts furrounding the broken bones, except clofe to the bones where the adhefive inflammation had taken place only in a fmall degree.

The bones were found to ride confiderably, viz. near three inches.

The cavity made in the foft parts, in confequence of the laceration made by the riding of the bones, had its parietes thickened, and pretty folid, by means of the adhefive inflammation, although not fo much as would have been the cafe, if the parts had been better difpofed for inflammation: and fome parts had become bony.   There hardly was found within this cavity any extravafated blood, or coagulating lymph except a few pretty loofe fibres like ftrings, which were vifibly the remains of the extravafated blood.

From thefe appearances this cavity had evidently loft its firft bond of union, viz. the extravafated blood, which took place from the ruptured veffels, and probably the fecond had never taken place, viz. the coagulating lymph, in confequence of the adhefive inflammation : however, there was an attempt towards an union, for the furrounding foft parts, we have obferved, had taken on the adhefive, and offific inflammation ; fo that in time there might have been formed in the furrounding foft parts a bony cafe, which would have united the two bones ; but the parts being deprived of the two common modes of union, they were led to a third.

From the ends of the bones, and fome parts of their furface, as well as from the inner furface of the foft parts, there was formed new flefh, fimilar to granulations.

The hollow ends of the bones were filled with this matter, which was rifing beyond the common furface of the bone ; and in fome places adhefion had taken place between it and the furrounding parts, with which it had come in contact. The same appearance, which this new flesh had in this case, I have several times seen in joints, both on the ends of the bones, and on the inside of the capsular ligament, but never before understood how it was formed: hence we find that granulations can, and do arise in parts that are not exposed. This is what I have long suspected to be the case in the union of the fractured patella, and this fact confirms me more in that opinion.

Here then we are shewn that, the cause of granulation, or the forming of new flesh for union (independent of extravasation, or the adhesive inflammation) is more extensive in its effects than we were formerly acquainted with; and that granulations, or new flesh, arise in all cases from the first and second bond of union being lost in the part, (which indeed seldom happens, except from exposure) it therefore makes no difference, whether the first and second bond of union escape through an opening made in the skin, as in a compound fracture, or it loofes its living powers, as in the present case, and as I suppose to be the case in a fracture of the patella, which obliges the absorbents to take it up as an extraneous body.

II. THE NATURE AND PROPERTIES OF GRANULATIONS.

GRANULATIONS, and this new formed substance, are an accretion of animal matter upon the wounded or exposed surface: they are formed by an exudation of the coagulating lymph from the vessels, into which new substance both the old vessels very probably extend, and also entirely new ones form, so that the granulations come to be very vascular, and indeed they are more so than almost any other animal substance. That this is the case, is seen in sores every day. I have often been able to trace the growth and vascularity of this new substance. I have seen upon a sore a white substance, exactly similar in every visible respect

to coagulating lymph. I have not attempted to wipe it off,
and the next day of dreffing I have found this very fubftance
vafcular; for by wiping or touching it with a prob , it
has bled freely. I have obferved the fame appearance on
the furface of a bone that has been laid bare. I once f -
ped off fome of the external furface of a bone of the fo t,
to fee if the furface would granulate. I remarked the following day that, the furface of the bone was covered with
a whitifh fubftance, having a tinge of blue; when I paffed
my probe into it, I did not feel the bone bare, but only
its refiftance. I conceived this fubftance to be coagulating
lymph, thrown out from inflammation, and that it would
be forced off when fuppuration came on; but on the fuc-
ceeding day I found it vafcular, and appearing like healthy
granulations.

The veffels of granulations pafs from the original parts,
whatever thefe are, to the bafis of the granulations; from
thence towards their external furface, in pretty regular
parallel lines, and would almoft appear to terminate there.

The furface of this new fubftance, or granulations, continues to have the fame difpofition for the fecretion of pus,
as the parts from which they were produced; it is therefore
reafonable to fuppofe that, the nature of the veffels does not
alter by forming the granulations; but that they were completely changed for the purpofe before the granulations began to form, and that thefe granulations are a confequence
of a change then produced upon them.

Their furfaces are very convex, the reverfe of ulceration,
having a great many points, or fmall eminences, fo as to
appear rough: and the fmaller thefe points are, the more
healthly we find the granulation.

The colour of healthly we find the granulations, is a deep
florid red, which would make us fufpect that the colour
was principally owing to the arterial blood*; but it only
fhews a brifk circulation in them, the blood not having time
to become dark.

When naturally of a livid red, they are commonly unhealthy and fhews a languid circulation, which appearance

---

* I once began to fufpect that the air might have fome influence
upon the blood, when circulating in the veffels, but from its
lofing that florid colour in fores of the legs by ftanding erect, I
gave up that idea.

often comes on in granulations of the limbs from the pofition of the body, as is evident from the following cafe.

A ftout, healthy, young man, had his leg confiderably
torn, and it formed a broad fore; when healing it was fome
days of a florid red, and on others of a purple hue: wonder-
ing what this could be owing to, he told me, when he
ftood for a few minutes it always changed from the fcarlet
to the modena.   I made him ftand up, and found it foon
changed: this plainly fhews that, thefe new formed veffels
were not able to fupport the increafed column of blood,
and to act upon it, which proves that a ftagnation of blood
was produced, fufficient to allow of the change in the colour, and moft probably both in the arteries and veins.

Thefe fores never heal fo faft as the others; whether it is
occafioned by the pofition of the body, or the nature of the
fore itfelf, but moft frequently fo in cafes of the laft-men-
tioned kind.   As the pofition of the body is capable of producing fuch an effect, it fhews us the reafon why fore legs
are fo backward in healing, when the perfon is allowed
to ftand or walk.

Granulations when healthy, and on an expofed or flat
furface, rife nearly even with the furface of the furrounding
fkin, and often a little higher; and in this ftate they are
always of a florid red; but when they exceed this, and take
on a growing difpofition, they are then unhealthy, become
foft, and fpongy, and without any difpofition to fkin.
Granulations are always of the fame difpofition with the
parts upon which they are formed, and take on the fame
mode of action.   If it is a difeafed part, they are difeafed;
and if the difeafe is of any fpecific kind, they are alfo of
the fame kind, and of courfe produce matter of the fame
kind, which I obferved when on pus.

Granulations have the difpofition to unite with one another when found, or healthy; the great intention of which
is, to produce the union of parts, fomewhat fimilar to that
by the firft intention, or the adhefive inflammation, although
poffibly not by the fame means.

The granulations having a difpofition to unite with each
other upon coming into contact, without the appearance of
any intermediate animal fubftance, perhaps is in the following
manner.   When two found granulations approach together, the mouths of the fecreting veffels of the one com-

ing to oppofe the mouths of fimilar veffels of the other, they are ftimulated into action, which is mutual; fo that a kind of fympathetic attraction takes place, and as they are folids, the attraction of cohefion is eftablifhed between them; this has been termed inofculation. The veffels thus joined, are altered from fecreting to circulating; or it may be in this way, viz. the circulatory veffels come to open upon the furface, and there unite with one another, and the two become one fubftance; or it may be afked, do they throw out coagulating lymph, when they come into contact, and have a difpofition to heal? and does this be-come vafcular, in which the veffels may inofculate, fimilar to union by the firft, or fecond intention? '

I have feen two granulations on the head, viz. one from the dura mater, (after trepanning) and the other from the fcalp, unite over the bare bone which was between them, fo ftrongly in twenty-four hours, that they required fome force to feparate them, and when feparated they bled.

The inner furface of the cutis in an abfcefs, or fore, does not only, not readily granulate, as has been mentioned, but it does not readily unite with the granulations underneath. The final intention of both feems to be, that the mouth of a fore which is feldom fo much in a difeafed ftate, fhould have a natural principle which attends difeafe, to put it upon a footing with the difeafe which is underneath; therefore, when abfceffes are allowed to become as thin as poffible be-fore they are opened, this proportion between the found fkin and the difeafe is better preferved, and the parts are not fo apt to turn fiftulous.

When the parts are unfound, and of courfe the granula-tions formed upon them unfound, we have not this difpofi-tion for union, but a fmooth furface is formed fomewhat fimilar to many natural internal furfaces of the body, and fuch as have no tendency to granulate; which continues to fecrete a matter expreffive of the fore which it lubricates, and in fome meafure prevents the union of the granulations. I imagine, for inftance, that the internal furface of a fiftu-lous ulcer is in fome degree fimilar to the inner furface of the urethra, when it is forming the difcharge commonly called a gleet. Such fores have therefore no difpofition in their granulations to unite, and nothing can produce an u-nion between them, but altering the difpofition of thefe granulations by exciting a confiderable inflammation, and probably ulceration, fo as to form new granulations, and by thefe means give them a chance of falling into a found ftate.

Granulations are not endowed with the fame powers as parts originally formed. In this refpect they are fimilar to all new formed-parts; and it is from this caufe that changes for the worfe are fo eafily effected. They more readily fall into ulceration, and mortification, than originally formed parts; and from their readinefs to ulcerate, they feparate floughs more quickly.

The granulations not only fhew the ftate of the part in which they are formed, or the ftate in which they are themfelves, but they fhew how far the conftitution is affected by many difeafes. The chief of thofe habits which affect the granulations in confequence of the conftitution, are, I believe, the indolent and irritable habits, but principally fevers; and thefe muft be fuch as produce univerfal irritation in the conftitution.

The unfound appearances of the granulations fhew to what a ftand the animal powers are put on fuch occafions, which does not appear fo vifibly in the originally formed parts; it is therefore evident, that the powers of the granulations are much weaker than thofe of the original parts.

III. LONGEVITY OF GRANULATIONS.

GRANULATIONS are not only weaker in performing the natural or common functions of the parts to which they belong, but they would appear often to be formed with only ftated periods of life, and thofe much fhorter than the life of the part on which they are formed. This is moft remarkable in the extremities; but where they are capable of going through all their operations, as cicatrization, their life then feems to be not fo limited : they are probably then acquiring new life, or longevity every day; but while in a ftate of granulation, we find them often dying without any vifible caufe: thus, a perfon fhall have a fore upon the leg, which fhall granulate readily, the granulations fhall appear healthy, the fkin fhall be forming round the edges, and all fhall be promifing well, when all at once the granulations shall become livid, lose their life, and immediately slough off ; or, in some cases, ulceration shall in part take place, and both together shall destroy the granulations ; and probably where ulceration wholly takes place, it may be owing to the same cause. New granulations shall immediately arise as before, and go through the same process ; this shall happen three or four times in the same person, and probably for ever, if some alteration in the nature of the parts be not produced. This circumstance of the difference in longevity of granulations in different people, is somewhat similar to the difference in longevity of different animals.

In cases of short lived granulations, I have tried various modes of treatment, both local and constitutional, to render the life of these granulations longer ; but without success.

It would appear from what has been said of suppuration and granulations, that it is absolutely necessary they should take place in wounds which are not allowed to unite by the first intention, before union and cicatrization can take place. Although this in general is the case, yet in small wounds, such as considerable scratches, or where there is a piece of skin rubbed off, we find that by the blood being suffered to coagulate upon the sore, and form a scab, which is allowed to remain, the sore will only be attended by the adhesive inflammation, and will skin over without ever suppurating ; where a small caustic has been applied, we find also, by allowing the slough to dry or scab, that when this is completed the scab will drop off, and the parts shall be skinned ; but if the blood has not been allowed to coagulate and dry, or the slough has been kept moist, the sore will suppurate and granulate.

We even see in small sores, which are perfectly healthy, and suppurating, that if the matter be allowed to dry upon them, the suppuration will stop, and the skin form under the scab ; the small-pox is a striking proof of this, which was fully treated of in a former part of the work.

A blister whose cuticle is not removed, similar to a scab. It does not allow of suppuration. If a separation takes place between the cutis and cuticle, and the cuticle be not removed, nothing will be collected through the whole course, and a new cuticle will be formed ; but if the cuticle be removed, a greater degree of inflammation will come on, and suppuration will certainly take place.

IV. OF THE CONTRACTION OF GRANULA-
TIONS.

IMMEDIATELY upon the formation of the granulations, cicatrization would appear to be in view. The parts which had receded, in confequence of a breach being made into them, by their natural elafticity, and probably by mufcular contraction, now begin to be brought together by this new fubftance ; and it being endowed with fuch properties, they foon begin to contract, which is a fign that cicatrization is to follow. The contraction takes place in every point, but principally from edge to edge, which brings the circumference of the fore towards the centre ; fo that the fore becomes fmaller and fmaller, although there is little or no new fkin formed.

The contracting tendency is in fome degree proportioned to the general healing difpofition of the fore, and the loofenefs of the parts on which they are formed ; for when it has not a tendency to fkin, the granulations do not fo readily contract, and therefore contracting and fkinning are probably effects of one caufe. The granulations too being formed upon a pretty fixed furface, which is a confequence of inflammation, are in fome degree retarded in their contraction from this caufe ; but probably this does not act fo much upon a mechanical principle as we at firft might imagine; for fuch a ftate of parts in fome degree leffens the difpofition for this procefs, but this ftate is every day altering, and in proportion as the tumefaction fubfides. Granulations are alfo retarded in their contraction, from a mechanical caufe, when they are formed on parts naturally fixed, fuch as a bone ; for inftance, on the fkull, the bone, etc. of the fhin, for there the granulations cannot greatly contract*.

In cafes where there has been a lofs of fubftance, making a hollow fore, and the contraction has begun, and advanced pretty far, before the granulations have had time to rife as high as the fkin, in fuch cafes the edges of the fkin are generally drawn down, and tucked in by it, in the hollow direction of the furface of the fore.

* This obfervation fhould direct us in operations on thofe parts, to fave as much fkin as poffible.

If it is a cavity, or abfcefs, which is granulating, with only a fmall opening, as in many that have not been freely opened, the whole circumfreence contracts, like the bladder of urine, till little or no cavity is left; and if any cavity is remaining; when they cannot contract any further, they unite with the oppofite granulations, in the manner above defcribed.

This contraction in the granulations continues till the whofe is healed, or fkinned over; but their greateft power is at the beginning, at leaft their greateft effect is at the beginning; one caufe of which is that, the refiftance to their contraction in the furrounding parts is then leaft.

The contractile power can be affifted by art, which is a further proof that there is a refiftance to be overcome.

The art generally made ufe of is that of bandages, which tend to pufh, draw, or keep the fkin near to the fore which is healing; but this affiftance need not be given, or is at leaft not fo neceffary, till the granulations are formed, and the contractile power has taken place: however, it may not be amifs to practice it from the very beginning, as by bringing the parts near to their natural pofition the adhefive inflammation will fix them there; they will therefore not recede fo much afterwards, and there will be lefs neceffity for the contractile powers of the granulations.

Befides the contractile powers of the granulations, there is alfo a fimilar power in the furrounding edge of the cicatrizing fkin, which affifts the contraction of the granulations, and is generally more confiderable than that of the granulations themfelves, drawing the mouth of the wound together like a purfe; this is frequently fo great, as to occafion the fkin to grafp the granulations which rife above the furface, and is very vifible in fugar-loaf ftumps, where the projection of the fore is to be confidered as above the level of the fkin.

This contractile power of the fkin is confined principally to the very edge where it is cicatrizing; and, I believe, is in thofe very granulations which have already cicatrized; for the natural, or original fkin furrounding this edge does not contract, or at leaft not nearly fo much, as appears by its being thrown into folds, and plaits, while the new fkin is fmooth and fhining. This circumftance of the original furrounding fkin not having the power of contraction, makes round wounds longer in healing than long ones;

for it is much eafier for the granulations, and the edge of the skin, to bring the fides of an oblong cavity together, than the fides of a circle ; the circumference of a circle not being capable of being brought to a point.

Whether this contraction of the granulations is owing to an approximation of all the parts, by their mufcular contraction, like that of a worm, while they lofe in fubftance as they contract ; or if they loofe without any mufcular contraction by the particules being abforbed, fo as to form interftices, (which I have called interftitial abforption) and the fides afterwards fall together, is not exactly determined, and perhaps both take place.

The ufes arifing from the contraction of the granulations are various. It facilitates the healing of a fore, as there are two operations going on at the fame time, viz. contraction and fkinning.

It avoids the formation of much new fkin, an effect, very evident in all fores which are healed, efpecially in found parts.

In amputation of a thick thigh (which is naturally feven, eight, or more inches diameter before the operation) the furface of the fore is of the fame diameter ; for the receding of the skin here does not increafe its furface, as it does in a cut on a plane ; yet in this cafe, he cicatrix fhall be no broader than a crown piece. This can be effected by the contractile power of the granulations, for it is bringing the skin within its natural bounds.

The advantage arifing from this is very evident, for it is with the skin, as it is with all other parts of the body, viz. that thofe parts which were originally formed are much fitter for the purpofes of life, than thofe that are newly formed, and not nearly fo liable to ulceration.

After the whole is skinned, we find that the fubftance which is the remains of the granulations on which the new skin is formed, ftill continues to contract, till hardly any thing more is left than what the new skin ftands upon. This is a very fmall part of the comparifon with the firft formed granulations, and it in time lofes moft of its apparent veffels, becomes white, and ligamentous. For we may obferve that, all new-healed fores are redder than the common skin, but in time they become much whiter.

As the granulations contract, the furrounding old skin is ftretched to cover the parts which had been deprived of

?:in, and this is at firſt little more than bringing the ſkin
to its old poſition, which had receded when the breach was
firſt made ; but afterwards it becomes conſiderably more,
ſo as to ſtretch, or oblige the old skin to elongate ; from
which we might ask the following queſtion :

Does the ſurrounding skin in the healing of a ſore length-
en by growth, or does it lengthen by ſtretching only ? I
think that the former is moſt probable ; and if this is the
caſe, I ſhould call this proceſs interſtitial growth, ſimilar to
the growth of the ears of the people in the Eaſtern iſlands,
particularly as it is an oppoſite effect to interſtitial abſorption.

Granulations appear to have other powers of action beſides ſimply their œconomy tending to a cure. They have
power of action in the whole, ſo as to produce other
operations, and even to affect other matter. I conceive that
a deep wound, ſuch as a gun-ſhot wound, advanced to ſuppuration, and granulation, and alſo a fiſtula, becomes in
ſome degree ſimilar to an execretory duct, having the powers of a periſtaltic motion from the bottom towards the o-
pening externally. Thus we find that whatever extraneous body is ſituated at the bottom of the ſore, is by degrees
conducted to the skin, although the bottom of the ſore, or
fiſtula, is of ſame depth. This effect in ſuch ſores does
not ariſe from the granulations forming at the bottom, and
gradually raiſing the extraneous body as they form, which
is commonly the caſe with exfoliations and ſloughs) but we
find extraneous bodies come to the skin when the bottom of
the wound is not granulating.

CHAPTER VIII

OF SKINNING.

WHEN a fore begins to heal, we find that the furrounding old skin, clofe to the granulations (which had been in a ftate of inflammation, having probably a red fhining furface, as if excoriated, and rather ragged) now becomes fmooth, and rounded with a whitifh caft as if covered with fomething white, and the nearer to the cicatrizing edge, the more white it is. This is, I believe, a beginning cuticle, which appearance is probably as early a fymptom of healing, and as much to be depended upon as any ; fo that the difpofition in the granulations for healing is manifefted in the furrounding skin ; and while the fore retains its red edge all round for perhaps a quarter, or half of an inch in breadth, we may be certain it is not a healing fore, and is what may be called, an irritable fore.

Skin is a very different fubftance, with refpect to texture, from the granulations upon which it is formed ; but whether it is an addition of new matter, viz. a new-formed fubftance upon the granulations being produced by them, or a change in the furface of the granulations themfelves, is not eafily determined. In either cafe, however, a change muft take place in the difpofition of the veffels, either to alter the ftructure of the granulations, or to form new parts upon them.

One would at firft be inclined to the former of thefe opinions, we have a clearer idea of the formation of a new fubftance, than fuch an alteration in the old. We find the new fkin moft commonly taking its rife from the furrounding old fkin, as if elongated from it ; but this is not always the cafe. In very large fores, but principally old ulcers, where the edges of the furrounding fkin have but little

D d

tendency to contract, or the cellular membrane underneath to yield, as well as the old skin having but little difposition to skinning in itself, a cicatrizing difposition cannot be communicated from it to the nearest granulations by continued sympathy. In such cafes new fkin forms in different parts of the ulcer, ftanding on the furface of the granulations, like little iflands. This, I believe, never takes place in parts the firft time of their being fore, nor in fores which have a ftrong propenfity to fkin.

Skinning is fomewhat like chryftalization, it requires a furface to fhoot from, and the edge of the fkin all round would appear to be this furface.

Whatever change the granulations undergo to form fkin they may in general be faid to be guided to it by the furrounding fkin, which gives this difposition to the furface of the adjoining granulations ; as adjacent bones give an offifying difposition to the granulations that are formed upon them. This may arife from fympathy ; and if it does, I fhould call it continued fympathy. But when the old fkin is unfound, and not able to communicate this difposition, then the granulations fometimes of themfelves acquire it, and new fkin begins to form where that difposition is ftrongeft in them, fo that the granulations may be ready to form new fkin, if the furrounding fkin be not in a condition to give the difposition. It would appear, however, that the circumference of the fore generally has the ftrongeft difposition to skin, even although the furrounding fkin does not affift ; for in many old fores no new fkin fhall fhoot from the furrounding fkin or be continued, as it were from the old ; and yet a circle of new skin fhall form, making a circle within the old, and as it were, detached from it.

Skinning is a procefs in which nature is always a great œconomift, without a fingle exception : this, however, may probably arife from granulations being always of the nature of the parts on which they are formed, and from feldom being formed on parts that are the leaft of the nature of the skin, they have therefore no ftrong difposition to form skin. What would feem to make this obfervation more probable, is, that if the cutis is only in part deftroyed, as by a hurt, or cauftic, which has not gone quite through the cutis to the cellular membrane underneath, a new cutis will form immediately on the granulations, and in many cafes it will form as faft as the flough will feparate ; the reafon is, becaufe the cutis has a ftronger tendency to form cutis than any other part, and in many cafes it may be faid to form it from almoft every point.

We never find that the new-formed fkin is fo large as the fore was, on which it is formed; this, I have already obferved, is brought about by the contraction of the granulations, which in fome meafure is in proportion to the quantity of furrounding old fkin, attended with the leaft refiftance.

If the fore is in a part where the furrounding fkin is loofe, as in the fcrotum, then the contractile power of the granulations being not at all prevented, but allowed full fcope, a very little new fkin is formed ; whereas, if the fore is on any other part, where the fkin is not loofe, fuch as the fcalp, fhin-bone, etc. in that cafe the new fkin is nearly as large at the fore.

This we find to be the cafe alfo in parts which are fo fwelled as to render the fkin tight, fuch as the fcrotum, when under the diftention of a hydrocele, and which fometimes happens where a cauftic has proved ineffectual ; we then find the new fkin as extenfive as in any other parts equally diftended. The fame thing takes place in white fwellings of the joint of the knee ; for if a fore is made upon fuch a part, as is frequently done by the application of cauftics, we find that the new fkin is nearly of the fame fize as the original fore. The general principle is alfo very obfervable after amputations of the limbs ; for if much old fkin has been faved, we find the cicatrix fmall, while on the other hand, if fuch care has not been taken, the cicatrix is proportionably large.

The new fkin is at firft commonly on the fame level with the old, and if there has not been much lofs of fubftance, or the difeafe is not very deep feated, it continues its pofition ; but this does not appear to be the cafe with fcalds and burns, for they frequently heal with a cicatrix, higher than the fkin, although the granulations have been kept even with the fkin. It would appear in thefe cafes that a tumefaction of the parts, which were the granulations, takes place after cicatrization.

Sometimes granulations cicatrize while higher than the common furrounding fkin, but when they are fuch as have been long in that pofition, as is the cafe in fome iffues : I have feen the granulations furrounding a pea rife confiderably above the fkin, near half-a-crown in breadth, and fkin

over, all but the hole in which the pea lay, the whole looking like a tumor.

I. THE NATURE OF THE NEW CUTIS.

THE new-formed cutis is neither so yielding nor so elaftic as the original is, and is also less moveable upon the
part to which it is attached, or upon which it is formed.
This last circumstance is owing to its basis being granulations, which are in some degree fixed upon parts united by
the adhesive inflammation ; and more particularly so, when
the granulations arise from a fixed part, such as a bone ;
the new skin formed upon them being also fixed in proportion.

It is, however, constantly becoming more and more flexible in itself, and likewise more loosly attached, owing to
the mechanical motion to which the parts are subject afterwards. The more flexible and loose the parts become, it
is so much the better, as flexibility, or the yielding of the
parts, preferves it from the effects of many accidents.
Parts which have been thickened in consequence of inflammation, such as the surrounding parts of new skin have always a less internal power of action in them, than parts
which have never been inflamed. This arises from the adventitious substance thrown out in the time of inflammation,
being a clog upon the operations of the original ; and the
new matter not being endowed with the same powers, the
part affected, taken as a whole, is by these means considerably weakened.

Motion given to the part so affected, must be mechanical ; but that motion becomes a stimulant to the parts moved, that they cannot exift under such motion without a-
dapting the structure of the parts to it, and this fets the abforbents to work, or they receive the stimulus of neceffity,
and abforb all the adventitious or rather superfluous fub-
stance ; by which means the parts are as much as possible
reduced to their original texture.

Medicines have not the powers we could wish in many
such cases; mercury, however, appears to have the power
of producing a similar stimulus to motion, and should be
made use of where a mechanical stimulus cannot be applied; and, I believe, when joined with camphire, its powers
of producing absorption are increased; when both medicine
and mechanical means can be used, so much the more benefit will ensue.

When every thing else fails, electricity might be tried.
It has been the cause of absorption of tumors. It has reduced the swellings of many joints in consequence of sprains,
and thereby allowed of the freedom of motion.

The new-formed cutis is at first very thin and extremely
tender, but afterwards becomes firmer and thicker: it is
a smooth continued skin, not formed with those insensible
indentations which are observed in the natural or original
skin, and by which the original admits of any distention
the cellular membrane will allow of, as is experienced in
many dropsies, white swellings in the joints, etc. This is
proved by steeping a piece of dead skin, with a cicatrix in
it, in water to make the cuticle separate from the cutis;
there we find that the new-formed cuticle becomes but little larger by such a process, which plainly shews, that the
new formed cutis upon which this cuticle was formed, has
a pretty smooth continued surface, and not that soft, une-
qual surface which distinguishes the original cutis.

This new cutis, and indeed all the substance which had
been formerly granulations, is not nearly so strong, nor endowed with such lasting and proper actions, as the originally formed parts. The living principle itself is also not nearly so active; for when an old sore once
breaks out, it continues to yield till almost the whole
of the new-formed matter has been absorbed or mortified; as has been already explained.

The young cutis is extremely full of vessels, which afterwards, in a great measure, either become lymphatic or
impervious, or taken into the constitution, so that the skin
and granulations underneath are at last free from visible
vessels and become white.

The surrounding original cutis, being drawn towards a centre by the contraction of granulations, to avoid as much as
possible the formation of new skin is thrown into loose folds,
while the new looks like a piece of skin upon the

ſtretch, and the whole appears as if a piece of skin had
been ſewed into a hole by much too large for it ; and
therefore it had been neceſſary to throw the ſurrouid-
ing old skin into folds, or gather the ſurrounding skin, in
order to bring it in contact with the new. The new cu-
tisof a ſore, I believe, never acquires a muſcular ſtructure;
nor does it grow larger than the ſore which it covers, ſo as
to be thrown into wrinkles ſimilar to the old ; and therefore has always that ſtretched, ſhining appearance.

II. OF THE NEW CUTICLE.

IT does not appear to be ſo difficult a proceſs for the cutis to form cuticle, as it is for the granulations to form
cutis ; for we find in general, that wherever there is a new
cutis formed, it is covered with a cuticle : and in caſes of
bliſters, or any other cauſe which may have deprived the
cutis of its cuticle, we find that the cuticle is ſoon reſtored.
We are to obſerve, however, that in ſuch caſes it is a ſound,
original cutis, forming its own cuticle, and having the
whole power of forming the cuticle, the ſurrounding cuticle itſelf having no power of action of this kind : every
point of cutis is forming cuticle, ſo that it is forming equally every where at once ; whereas I obſerved that, the formation of the cutis was principally progreſſive from the
ſurrounding cutis.

It is at firſt very thin, and partakes more of a pulpy than
a horny ſubſtance ; as it gets ſtronger, it becomes ſmooth
and ſhining, and is much more tranſparent than original cuticle, which ſhews more the colour of the rete mucoſum.
This account relates to the cuticle of ſound parts which
had gone through all the operatiohs of health, but where
there is a retardation in the healing we find that the cuticle
is, in ſome caſes, backward in forming, and in others it
ſhall be formed very thick, ſo as to make it neceſſary to be
removed, it appearing to be a clog upon the cutis, retard-
ing the progreſs of its formation.

III. OF THE RETE MUCOSUM.

The rete mucofum is later in forming than the cuticle, and in fome cafes never forms at all : this is beft known in blacks, who have been either wounded or bliftered, for the cicatrix in the black is a confiderable time before it becomes dark ; and in one black who came under my obfervation, a fore which had been upon his leg when young, remained white when he was old. After blifters too, the part bliftered remains white for fome time after the cuticle is completely formed : however, in many cicatrices of blacks, we find them even darker than any other part of the fkin.

CHAPTER IX

EFFECTS OF INFLAMMATION, AND ITS CONSEQUENCES ON THE CON- STITUTION.

THE conſtitutional affections ariſing from inflammation, are immediate, and remote.

The immediate affections have been already conſidered, viz. the ſympathetic fever, and alſo the nervous.   I ſhall now treat of the remote, viz. the hectic, and diſſolution, which ariſe from the ſtate of the local affection at the time ; the inflammation not being able to go through all the ſalu- tary ſteps that have been deſcribed.   We have diſeaſes, however, ſometimes accompanying thoſe ſalutary proceſſ- es, although we ſhould naturally conclude, from the fore- going account, that the ſuppurative inflammation and ſup- puration itſelf ſhould produce no change in the conſtitu- tion, but what was attendant upon the inflammation,   and might be ſuppoſed, perhaps, ſomewhat neceſſary to it ; and that when inflammation had ſubſided, and a kind of ſup- puration come on, the conſtitution ſhould be left in a ſound ſtate, becauſe it would now appear that all the future proceſſes were ſettled; and a conſtitution that was capable of doing this, was alſo capable of going through all the ſuc- ceeding operations, as they are only actions of reſtoration ; but we find ſometimes the contrary, and the condition in which the conſtitution is either left, or which it afterwards takes on, proves often much more hurtful than the inflam- mation itſelf.

Vol. II.   E e

It appears in many cafes that, the inflammation, the attendant fever, the going off of thefe, and the commencement and continuance of the fuppuration, produce in many perfons a change in the conftitution, giving a difpofition to fymptoms, which are called nervous. The locked jaw is often the effect of this leading caufe, as well as the hyfterics, fpafms upon the mufcles of refpiration, and great reftleffnefs, which often prove fatal to the patient ; there are, likewife, figns of great and univerfal debility, or figns of diffolution in the patient, all of which appear to be increafed by a continuance of the fuppuration. Each of thefe difeafes are well marked, and it would appear that the locked jaw, hifterics, fpafms, and great reftleffnefs, are of the nervous kind, and do not appear to arife from fuch a conftitution, as is not equal to overcome the caufe ; for the caufe which produced them being removed, the effects are going on towards health now; as well as before ; and if the patient dies of any of thofe difeafes, it is not from the caufe, nor from the immediate effect, viz. the local difeafe, but from the effect which the preceding operations, joined with the healing, have on fome conftitutions. They all feem to derive their origin from the fame root, viz. from all the foregoing proceffes, which we have been defcribing; but they are altogether too extenfive for our prefent fubject.

I. OF THE HECTIC.

I HAVE now defcribed the injuries of which inflammation is a confequence ; the progrefs of that action in different parts ; its effects on the conftitution ; together with the mode of treatment of both, and have carried it through its various fteps to a perfect reftoration. I have alfo already mentioned, that the act of abforption affects fome conftitutions ; but I fhall now take notice, that nature is not always equal to thofe falutary proceffes, and hence the conftitution fometimes becomes particularly affected, producing fymptoms different from thofe formerly defcribed, and which have been called the hectic.

This difeafe is one of our remote conftitutional fympathetic affections, and appears to arife from a very different
origin, from the other fympathizing effects beforementioned. When it is a confequence of a local difeafe, it has
commonly been preceded by the firft procefs of the former, viz. inflammation and fuppuration, but has not been
able to accomplifh granulation and cicatrization: fo as to
complete the cure. It may be faid to be a conftitution now
become affected with a local difeafe or irritation, which
the conftitution is confcious of, and of which it cannot
relieve itfelf, and cannot cure; for while the inflammation
lafts, which is only preparatory, and an immediate effect of
moft injuries, and in parts which can only affect the conftitution, fo as to call up its powers, there can be no hectic.

We fhould diftinguifh well between a hectic arifing from
a local complaint entirely, where the conftitution is good,
but only difturbed by too great an irritation; and a hectic
arifing principally from the badnefs of the conftitution,
which does not difpofe the parts for a healing ftate; for
in the firft it is only neceffary to remove the part (if remo-
vable) and then all will do well; but in the other we gain
nothing by a removal, except the wound made by the operation is much lefs, and much more eafily put into a local
method of cure; fo that this bad conftitution falls lefs under this, (the operation taken into the account) than under the former ftate; but all this depends on nice difcrimination.

The hectic comes on at very different periods after the
inflammation, and commencement of fuppuration, owing
to a variety of circumftances. Firft, fome conftitutions
much more eafily fall into this ftate than others, having
lefs powers of refiftance. The quantity of incurable difeafe muft be fuch as can affect the conftitution, and in
whatever fituation, or in whatever parts, it will be always
as to the quantity of difeafe in thofe fituations or parts in
the conftitution, which will make the time to vary very
confiderably. In many difeafes it would appear, from the
manner of coming on, that they retard the commencement
of the hectic, fuch as lumbar abfceffes. But when fuch
abfceffes are put into that ftate, in which the conftitution
is to make its efforts towards a cure, but is not equal to the
tafk, then the hectic commences.

It takes its rife from a variety of caufes, but which ! fhall divide into two fpecies, with regard to difeafed par* viz. the parts vital, and the parts not vital. The only diffe-rence between thefe two, is, probably, merely in time, with refpect to its coming on, and its progrefs when come on : but what is very fimilar to the difeafe of a vital part, is quantity of incurable difeafe.

The caufes of hectic, arifing from difeafes of the vital parts, may be many, of which a great proportion would not produce the hectic if they were in any other part of the body ; fuch, for inftance, as the formation of tumors, either in, or fo as to prefs upon fome vital part, or a part whofe functions are immediately connected with life. Schirri in the ftomach, mefenteric glands, which tumors any where elfe would not produce the hectic ; ma-ny complaints too of vital parts, as difeafed lungs, liver, etc. all of thefe produce the hectic, and much fooner than if the parts were not vital. In many cafes where thofe caufes of the hectic come on quickly, it frequently follows fo quick upon the fympathetic fever, that the one feems to run into the other : this I have often feen in the lumbar abfcefs. They alfo produce fymptoms according to the nature of the part injured, as coughs, when in the lungs ; ficknefs and vomiting, when in the ftomach ; and probably bring on many other complaints, as dropfies, jaundice, etc. but which are not peculiar to the hectic.

When the hectic arifes from a difeafe in a part not vital it fooner or later commences, according as it is in the pow-er of the parts to heal, or continue the difeafe. If far from the fource of the circulation, with the fame quantity of dif-eafe, it will come on fooner. When in parts not vital, it is generally in thofe parts where fo great a quantity of dif-eafe can take place, (without the power of being diminifh-ed in fize, as is the cafe with the difeafes in moft joints*) as to affect the conftitution, and alfo in fuch parts as have naturally but little powers to heal ; we muft at the fame time include parts that are well-difpofed to take on fuch fpecific difeafes as are not readily cured in any fituation ; fuch parts are principally the larger joints, both of the

---

* The cavity of a joint is fuch, as not readily to become fmaller under difeafe. as in the foft parts, which was defcribed in the contraction of fores.

trunk and extremities ; but in the fmall joints of the toes, and fingers, although the fame local effects take place, as in the larger, yet the conftitution is not made fenfible of it ; we therefore find a fcrofulous joint of a toe or finger going on for years, without affecting the conftitution.

The ankle, wrift, elbow, and even the fhoulder, may be affected much longer than either the knee, hip-joint, or loins, before the conftitution fympathizes with their want of powers to heal.

Although the hectic commonly arifes from fome incurable local difeafe of a vital part, or of a common part when of fome magnitude, yet it is poffible for it to be an original difeafe in the conftitution : the conftitution may fall into the fame mode of action, without any local caufe whatever, at leaft that we know of.

Hectic may be faid to be a flow mode of diffolution ; the general fymptoms are thofe of a low, or flow fever, attended with weaknefs, but more with the action of weaknefs than real weaknefs ; for, upon the removal of the hectic caufe, the action of ftrength is immediately produced, as well as every natural function, however much it was decreafed before.

The particular fymptoms are debility ; a fmall, quick, and fharp pulfe ; the blood forfaking the fkin ; lofs of appetite ; often rejection of all aliment by the ftomach ; wafting ; a great readinefs to be thrown into fweats ; fweating fpontaneoufly when in bed ; frequently a conftitutional purging ; the water clear.

This difeafe has been, and is ftill in general laid to the charge of the abforption of pus into the conftitution from a fore ; but I have long imagined that an abforption of pus has been too much blamed as the caufe of many of the bad fymptoms which frequently attack people who have fores.

Firft, this fymptom almoft conftantly attends fuppuration when in particular parts, fuch as the vital parts, as well as many inflammations before actual fuppuration has taken place, as in many of the larger joints, called white fwellings ; while the fame kind and quantity of inflammation and fuppuration in any of the flefhy parts, and efpecially fuch of them as are near the fource of the circulation, have in general no fuch effect ; in thofe cafes, therefore, it is only an effect upon the conftitution produced by a local complaint, having a peculiar property, which I fhall now confider.

I obferved, that with all difeafes of vital parts, the conftitution fympathized more readily than with difeafes of any other parts; and alfo, that all difeafes of vital parts are more difficult of cure in general than thofe which are not vital.   I have obferved, likewife that all the difeafes of bones, ligaments, and tendons, affected the conftitution more readily than thofe of mufcles, fkin, cellular membrane, etc. and we find that the fame general principles are followed in the univerfal remote fympathy, produced by local difeafes of thofe parts.

When the difeafe is in vital parts, and is fuch as not to kill by its firft conftitutional effects, the conftitution then becomes teazed with a complaint which is difturbing the neceffary actions of health, the parts being vital; there is, befides, the univerfal fympathy, with a difeafe which gives the irritation of being incurable.

In the large joints it continues to harrafs the conftitution with a difeafe, where the parts have no power, or what is more probable, have no difpofition to produce a falutary inflammation and fuppuration; the conftitution, therefore, is alfo irritated with an incurable difeafe.

This is the theory of the caufe of the hectic, which will be further illuftrated : but now let us confider how far the idea of the abforption of matter may be a caufe.

If the abforption of matter always produced fuch fymptoms, I do not fee how any patient, who has a large fore, could poffibly efcape this difeafe ; becaufe we have as yet no reafon to fuppofe, that any one fore has more power of abforption than another.

If in thofe cafes where there is an hectic conftitution, the abforption is really greater than when the habit is healthy, it will be difficult to determine whether this increafe of abforption is a caufe, or an effect.

If it be a caufe, it muft arife from a particular difpofition in the fore to abforb more at one time than common, even while it was in a healthy ftate; for the fore muft be healthy and then abforb, which hurts the conftitution ; moreover, as the fore is a part of that conftitution, it muft of courfe be affected in turn ; and what reafon we have to fuppofe that a healthy fore of a healthy conftitution fhould begin to abforb more at one time than another, I muft own I can-

not difcover. If this increafe of abforption does not depend upon the nature of the fore, it muft then take
its rife from the conftitution ; and if fo, there is then a
peculiarity in the conftitution, fo that the whole of the
fymptoms cannot arife entirely from the abforption of matter
as a caufe, but muft depend on a peculiar conftitution,
and abforption combined.

If abforption of matter produced fuch violent effects as
are commonly afcribed to it, (which indeed are never of
the inflammatory kind, but of the hectic) why does not the
venereal matter do the fame ? We often know that abforption is going on by the progrefs of buboes ; and I have
known a large bubo, which was juft ready to break, abforbed from a few days ficknefs at fea, while the perfon continued at fea for twenty-four days after ; yet, in fuch cafes,
no fymptoms appear till the matter begins to have its fpe-
fic effects, and thefe very fymptoms, are not fimilar to
thofe which are called hectic. From reafoning, we ought
to expect that the venereal matter would act with greater
violence than the common matter from a healthy fore.
Although matter too is frequently formed on the infide of
the veins, in cafes of inflammation of their cavities*, and
this matter cannot fail of getting into the circulation, yet in
thefe cafes we have not the hectic difpofition but only the
inflammatory, and fometimes death. We likewife find very large collections of matter, which have been produced,
without vifible inflammation, fuch as many of the fcrofulous
kind, and which are wholly abforbed, even in a very fhort
time, yet no bad fymptoms follow†.

We may, therefore, from hence ! conclude, that the abforption of pus from a fore into the circulation, cannot be
a caufe of fo much mifchief as is generally fuppofed ; and
if it was owing to matter in the conftitution, I do not fee
how thefe fymptoms could ever ceafe, till fuppuration ceafed, which does not readily happen in fuch conftitutions, their
fores being tedious in healing. We find, however, that
fuch patients often get well of the hectic before fuppura-

* Vide Tranfactions of a Society for the improvement of
medical and chirurgical Knowledge.
. † It may, however, be objected to this, that this is not
true matter, or pus ; but it may be neceffary to fhew that the
one affects the conftitution upon abforption more than the other.

tion ceafes, even when no medicine was given ; and in the
cafe of veins, there is great reafon to believe, that after all
the bad fymptoms are removed, fuppuration is ftill going on,
as we find it fo in a fore ; pus may, therefore, ftill pafs
into the conftitution from the veins, and yet the hectic may
not be produced, which would certainly be the cafe if thofe
bad fymptoms were occafioned by the matter getting into
the circulation.

But I very much doubt the fact of abforption going on
more in one fore than another; and if ever it does I think
it is of no confequence ; I am much more inclined to believe, that this hectic difpofition arifes from the effect
which irritation of a vital organ, and fome other parts, fuch
as joints, (being either incurable in themfelves, or being fo
to the conftitution for a time) have on the conftitution.

We may remark, that in large abfceffes which have not
been preceded by inflammation, the hectic difpofition feldom or never comes on till after they are opened, (although
they may have been forming matter for months) ; but in
fuch cafes, the difpofition often comes on foon after opening, and in others very late. Till the ftimulus for reftor-
ing parts is given, no fuch effect can take place ; and if
the parts are well-difpofed to heal no hectic difpofition
comes on, neither is the conftitution at all affected. In
difeafed joints alfo, which are attended with inflammation
if the parts were capable of taking on a falutary inflammation, we fhould have only the firft fympathetic fever ; but
as they feldom are capable of doing this, the conftitution
becomes teazed with a complaint, not taking on the immediate and falutary fteps towards a cure. In the venereal difeafe too, where we know that the venereal matter has got
into the conftitution, and that the matter is producing its
fpecific effects, yet no hectic comes on, till the conftitution is harraffed with an incurable difeafe, and this not till
long after all the parts are healed, with regard to recent difeafe, and no matter is formed for further abforption. That
abforption does not take place in fores, we have reafon to
believe, and upon this fact a mode of dreffing fores has been
advifed. The following is a remarkable inftance of it in
a bubo : a young man had a chancre and three buboes, one
of which appeared when the other two were almoft cured.
This was very large, and at the bottom of the belly. When
it had fuppurated, and was pretty near breaking, it dimi-

niſhed very quickly, and in two or three days was entirely
gone. While this was going on, he obferved his urine
wheyiſh and thick, while making it, which went entirely off when the bubo had ſubſided. Before the bubo
began to ſubſide, he was rather mending in his health,
which continued to mend, nor did the diminution of
the bubo alter the ſtate of his health.

The hectic, from what has been ſaid, appears in ſome
meaſure to depend on the parts being ſtimulated to produce
an effect which is beyond their powers: that this ſtimulus
is ſooner or later in taking place in different caſes, and that
the conſtitution becomes affected by it. The hectic diſpoſition ariſes from diſeaſed lungs, lumbar abſceſſes, white
ſwellings, ſcrofulous joints, etc.

II. THE TREATMENT OF THE HECTIC.

WE have as yet, I am afraid, no cure for any of the
conſequences above related; I believe that depends in
the cure of the cauſe, viz. the local complaint, or in its
removal; the effects, I fear, are not to be cured.
Strengtheners, and what are called antiſeptics, are recommended.

Strengtheners are propoſed on account of the debility
which has taken place.

Antiſeptics have been employed from an idea that pus,
when abſorbed, gives the blood a tendency to putrefaction.
To prevent both of theſe effects from taking place, the
ſame medicines are however recommended. Theſe are
bark and wine.

Bark will, in moſt caſes, only aſſiſt in ſupporting a conſtitution. I ſhould ſuppoſe it impoſſible to cure a diſeaſe of
the conſtitution till the cauſe be removed; however, it may
be ſuppoſed that theſe medicines may make the conſtitution leſs ſuſceptible of the diſeaſe, and may alſo contribute
to leſſen the cauſe, by diſpoſing the local complaints to heal:

but where the hectic arifes from fpecific difeafe ; as for inftance, if a hectic difpofition comes on from a \ ne-real difpofition, bark will enable the conftitution to fupport it better than it otherwife could have done ; but can never remove it.

Wine, I am fearful, rather does harm if it increafes the actions of the machine without giving ftrength, a thing carefully to be avoided ; however, I have not yet made up my mind about wine.

When the hectic arifes from local difeafes, in fuch parts as the conftitution can bear a removal of, then the difeaf-ed part fhould be removed, viz. when it arifes from fome incurable difeafe in an extremity, and although all the fymptoms above-defcribed fhould have already taken place, we fhall find that upon a removal of the limb the fymptoms will abate almoft immediately. I have known a hectic pulfe at one hundred and twenty fink to ninety in a few hours, upon the removal of the hectic caufe. I have known perfons fleep found the firft night without an opi-ate, who had not flept tolerably for weeks before. I have known cold fweats ftop immediately, as well as thofe call-ed colliquitive. I have known a purging immediately ftop, upon the removal of the hectic caufe, and the urine drop its fediment. It is poffible too, that the pain in the opera-tion, and the fympathetic affection of the conftitution may affift in thefe falutary effects. It is an action diametrically oppofite to the hectic, and may be faid to bring back the conftitution to a natural ftate.

III. OF DISSOLUTION.

DISSOLUTION is the laft ftage of all, and is common to, or an immediate confequence of all difeafes, whether local or conftitutional. A man fhall not recover of a fe-ver, whether original or fympathetic, but fhall move into the laft ftage, or diffolution. It fhall take place in the fe-cond ftage of a difeafe, where the ftate of conftitution

and parts appears to be formed out of the firft; as for in -
ftance, a man fhall lofe his leg, efpecially if above the knee;
or have a very bad compound fracture in the leg; the firft
conftitutional fymptoms fhall have been violent, but all
fhall appear to have been got the better of, and there fhall
be hopes of recovery, when fuddenly he fhall be attacked
with a fhivering fit, which fhall not perform all its actions,
viz. fhall not produce the hot fit and fweat, but fhall continue a kind of irregular hot fit, attended with lofs of
appetite, quick, low pulfe, eyes funk, and the perfon fhall
die in a few days. Or he fhall go into the common difeafed
fymptoms of the fecond ftage, viz. the nervous, with many
of its effects, as the tetanus, and diffolution fhall alfo be a
confequence. Or if the local difeafe does not or cannot heal,
and is fuch as to affect the conftitution, it then brings on
the hectic, and fooner or later diffolution takes place; for
the hectic is an action of difeafe, and of a particular kind;
but diffolution is giving way to difeafe of every kind, therefore has no determined form arifing from the nature of the
preceding difeafe.

It has been fuppofed, that this difeafe arifes alfo from
the abforption of matter. It appears to be in many cafes
an effect arifing from violent and long continued inflammations and fuppurations, although not incurable in themfelves; (therefore, in thofe refpects, not fimilar to the hectic) and which in many inftances are known to produce
the greateft changes in the conftitution. Such often arife
from very bad compound fractures, from amputations of
the extremities, efpecially the lower, and more particularly
the thigh, in which cafes the fympathetic fever has run high,
which would appear to be neceffary, or preparatory; but in
the hectic, it is not neceffary that the conftitution fhould
have fuffered at all in the firft ftages of the difeafe; diffolution feems to be more connected with what is paft, than
with the prefent alone, which is the reverfe of the hectic.
We never find this difeafe take place in confequence of fmall
wounds, or fuch wounds as have affected the conftitution
but little in its firft ftages; but which may affect the conftitution much in its fecond, fuch as fmall wounds producing the lock jaw. It would appear to take place in our
hofpitals more generally than in private houfes, and more
readily in large cities than in the country. We fhall find
that the hectic and this are by no means the fame difeafe,
differing exceedingly in their caufes, and in many of their

effects; for in the cafes of compound fractures and amputations, we find the conftitution often capable of going
through the inflammatory and fympathetic fever, producing fuppuration and granulation, as well as continuing the
production of thefe for fome time, yet finking under them
at laft, and often immediately, without a feeming caufe.
This effect will more readily take place, if the perfon was
in full health before the accident or operation, than if he
had been fomewhat accuftomed to the other, or true hectic;
for the fymptoms of diffolution feldom or never take place,
if the violence committed has been to get rid of a hectic caufe.
It fometimes takes place early, in confequence of local injury, and would feem to be a continuation of the fympathetic fever; as if the conftitution was not able to relieve
itfelf of the general affection, or that the parts could not
go into the true fuppurative difpofition.   We fee this frequently after removing a limb, efpecially in the lower extremity, and after cutting for the ftone in very fat men, a-
bove the middle age, and who have lived well.

   The firft fymptoms are generally thofe of the ftomach,
which produce fhivering: vomiting immediately follows, if
not an immediate attendant; there is great oppreffion and
anxiety, the perfons conceiving they muft die.   There is a
fmall quick pulfe; perhaps bleeding from the whole furface of the fore, often mortification with every fign of diffolution in the countenance; as it arifes with the fymptoms
of death, its termination is pretty quick.   Here is a very
fatal difeafe taking place; in fome almoft immediately,
when all appeared to be within the power of the machine,
and therefore cannot immediately arife from the fore itfelf;
for it is very common after fuch operations as ufually do
well; but the hectic always takes place in confequence of
thofe fores which feldom or never get well in any cafe; yet
the fore certainly affifts in bringing on diffolution, becaufe
we never fee the difeafe take place when the fore is healed, nor in thofe where the conftitution feems not to be
equal to the tafk, as is the caufe of the hectic.

   The hectic is much flower in its progrefs, and feems to
be a fimple and an immediate effect, arifing from a continued caufe which is local; by removing the caufe, therefore, the effect ceafes, and the havoc made upon the conftitution is foon reftored; perfons, therefore, do much better
in confequence of the hectic having in fome degree taken
place, prior to the removal of the caufe.   But diffolution

is a change of the conftitution in confequence of caufes
which now do not wholly exift, and in many cafes it does
not take place till the conftitution appears to be capable eafily of performing all its functions, and a removal of the
parts does not relieve, as in the hectic; for diffolution does
not depend for its continuance upon the prefence of the difeafe.

Death or diffolution, appears not to be going on equally
faft in every vital part; for we fhall have many people very
near their termination, yet fome vital actions fhall be good,
and tolerably ftrong; and if it is a vifible action, and life
depends much upon this action, the patients fhall not appear to be fo near their end as they really are: thus I have
feen dying people whofe pulfe was full and ftrong as ufual,
on the day previous to their death, but it has funk almoft
at once, and then become extremely quick, with a thrill:
on fuch occafions it fhall rife again, making a ftrong effort,
and after a fhort time, a moifture fhall probably come on
the fkin, which fhall in this ftate of pulfe be warm; but u-
pon the finking of the pulfe, fhall become cold and clammy:
breathing fhall become very imperfect, almoft like fhort
catchings, and the perfon fhall foon die.

It would appear in many cafes, that difeafe has produced fuch weaknefs as at laft, to deftroy itfelf: we fhall even
fee the fymptoms, or confequences of difeafe, get well before
death. A gentlewoman, who was above feventy-five, was
anafarcous allover: the abdomen was very full and large;
fhe made but very little water; her breathing was fo difficult as to make her purple in the face, fo that moft probably there was water in the cheft; her pulfe was extremely
irregular; fluttering, trembling, intermitting and fmall.
Her legs were punctured with a lancet, and difcharged
very freely for more than three weeks, which emptied the
cellular membrane of the body, as well as in fome degree
the abdomen; the breathing became free and eafy, fo that
we fuppofed the water in the cheft was abforbed; the pulfe
became regular, foft, and fuller, and the appetite in fome
degree mended; in which ftate fhe feemed free from difeafe, having only fome of the confequences ftill remaining.
The quantity of urine increafed to the natural quantity;
but notwithftanding actual difeafe feemed to be gone, yet
fhe became weaker and weaker, in which ftate fhe exifted
for near a month, and died. Some days prior to death, a

purple and then a livid appearance came upon the legs, with some spots of extravasated blood above where the punctures had been made, on which blisters arose, at first filled with serum, then with bloody serum, all of them threatening mortification.

Even when in the state of approaching death, we often find a soft, quiet, and regular pulse, having not the least degree of irritability in it, and this when there is every other sign of approaching death; such as entire loss of appetite, no rest, hickup, the feet cold and partial, cold, clammy sweats, etc.

A lady appeared to have lost all diseased action, only the consequences of disease remaining, viz. weakness, with swelled legs; she made little or no water; at length she became so weak, as hardly to articulate; she lay in a kind of doze, was only roused to impression, and only took food by spoonfuls when desired; the pulse so small as hardly to be felt: her extremities were cold, and she had all the signs of approaching dissolution, which took place; yet within thirty-six hours before she died, the whole water in her legs and thighs was taken up, her urine increased, and about ten hours previous to her death, the legs, etc. were as small as ever. As I consider the dropsy to be a disease, and not simply weakness, which this case would in some measure shew from the result, I should wish to ask, whether the absorption of water was not owing to the disease being gone, and whether the disease being gone, the absorbents did not set to work? If so, then dissolution may be a cessation of disease, and persons die of weakness simply; or simply, either the want of powers to act, or the want of that stimulus of necessity to act, by which means a cessation of action takes place.

Since bodies of persons who die suddenly, and even by violent death, as well as those who die soon after a considerable operation, are not capable of being preserved so long as those who have been ill for some time; and as those who have a considerable operation performed upon them, as the amputation of a leg, do not so readily recover as those who have been long ill, may not the more ready production of death, and the more ready production of putrefaction be owing to the same principle? one more readily running into the action of death, as also more readily into the action of putrefaction; but it is very probable that the action producing quick putrefaction, is an action prior to absolute death.

PART III.

CHAPTER I.

THE TREATMENT OF ABSCESSES.

═══════════════════════

I HAVE endeavoured to lay down the general principles of fuppuration, which principles of themfelves lead to
a general method of cure ; but as it is only the proper application of art, to thofe principles which completes the
furgeon, and fince it is the moft difficult part to apply our
knowledge of the firft principles to practice with readinefs,
efpecially when there appear fome peculiarities, it will be
neceffary to bring the beginner from firft principles to the
practical part.

Abfceffes are in general confequences of fpontaneous inflammation, but not always fo ; for they may be confequences
of fome violence, as ftrains or bruifes from fome external
violence, which has hurt deeper feated parts than the fkin
over them, which inflame and form an abfcefs, as was defcribed in treating of accidents ; as alfo from the introduction of extraneous bodies, over which the parts have healed. Even when they appear to be fpontaneous, they arife
from fo many caufes, and from thence have fo many difpofitions, or are of fo many kinds, that in general they become one of the greateft objects in furgery ; becaufe, from
thefe circumftances, they require a vaft variety in the manner of treatment.

I do not mean at present to enter into a full discussion of the cause, effect, and cure of every abscess, because that would be treating of every disease which is capable of producing such complaints, many of which would come under the article of specific diseases, which must be treated of separately ; yet I mean here to lay down such general surgical rules for their treatment and many of their consequences, as will include almost every kind of disease of this kind, considered as an abscess simply ; so that the specific treatment of any specific abscess will be principally confined to the medicinal treatment of the part and the constitution ; thereby the treatment of the local complaint so produced, abstracted from the specific disposition, will for the most part come under our general rules.

As most spontaneous suppurations, from whatever cause, are deeper seated than the surface of the body, such of course must form what are called abscesses, or collections of pus ; therefore we have abscesses of all depths, from the pimple in the skin to the boil ; and from the boil to deep-seated abscesses, among the muscles, or in any other deep-seated part.

Abscesses are commonly formed where matter is found, especially the more superficial ones, and such may be justly called abscesses of this part ; but collections of matter are often found in parts where not formed, more especially in the deeper seated ones, the matter moving from the seat where it was formed to some more depending part, or having met with some obstruction in its course, it takes another direction and therefore may be called an abscess in this part ; and I shall call them so in my descriptions of them ; I believe such abscesses do not arise from inflammation, but are of the scrofulous kind, and therefore not so much to our present purpose.

It will be difficult to divide abscesses into absolutely distinct classes ; but, similar to inflammation, they may be divided into two kinds, the sound and the unsound ; for I imagine these two first principles might lead to the method of cure ; but at present I only mean to lay down the principles of an abscess.

The appearances which distinguish the sound from the unsound abscess are several ; although there are many abscesses of particular kinds that give little or no information. They often differ from one another in their first appearance, from the kind of inflammation, as alfo in their courfe, but more particularly in their efforts towards a cure.

Thus we judge of the confequences of the fmall pox, from the firft appearance of the arm after innoculation ; for if the beginning inflammation is fmall, pretty much cir- cumfcribed, and of the florid red with fome rifing, then we may in our own minds expect a good kind ; the fame upon the firft appearance of the fmall pox themfelves ; as alfo the firft appearance of a chancre, etc. or almoft of a- ny other difeafe, either beginning with, or attended by in- flammation ; for it is by the kind of inflammation we are to judge of the future event.

It might be thought almoft unneceffary here to treat of found abfceffes, becaufe in fuch our firft principles will readily take place, and often little or no affiftance is requir- ed ; but abfceffes may be attended with circumftances which may retard the cure, and which have nothing to do with unfoundnefs ; fuch as extraneous bodies in found parts and thefe will moft probably come under our general principles of cure ; that is, require fomething to be done, becaufe they will in many cafes, relieve themfelves of the extraneous matter, and then they require but little affift- ance.

I. THE PROGRESS OF ABSCESSES TO THE SKIN.

WHAT I mean by a found abfcefs, is, where there is a found conftitution, the parts affected having all the difpo- fition and powers to heal ; thofe difpofitions and powers allowed to take place, which will take place more readily if in ftructures of the body which have naturally a ready difpofition to heal ; fo fituated in the body as to be able to fupport its actions, and not of a fpecific kind, for which we have no cure; for any specific disease, for which we have a cure, will come within our first division*.

The inflammation in a sound and active part, and of a sound constitution, in general is pretty violent, attended from the very beginning with a considerable deal of pain †, suppuration takes place quickly ; the parts between the abscess and the skin are readily affected, and ulceration goes on fast, the skin becomes of the florid red, the matter comes soon to it, especially at a point*, and it bursts ; all this is done with great rapidity

These symptoms show such a degree of health in the constitution and the parts, that little is necessary for the surgeon to do in the first stages of the disease.

Poultices are recommended in such cases to assist that disposition which the parts have to give way between the skin and the abscess ; but I have already observed, that they certainly can have no effect of this kind ; however, they have their uses when the inflammation has reached the skin, for they keep it soft, allow the cuticle to distend, and give way to the swelling underneath ; which eases the patient ; warmth and moisture act in many cases as sedatives to our sensations, although not always ; and the distinction between those where they give ease, and where they rather give pain, I have not been able to make out.

---

* Viz. If a venereal abscess has its specific quality destroyed, it admits of cure as readily as any other, and the same treatment becomes necessary.

† Vide symptoms of suppurative inflammation.

* This very appearance makes a material difference between an abscess arising from brisk inflammation, and one that is slow in its progress ; it is so remarkable, that I have seen this effect where the matter was at such distance as not to be felt in the least, and where I have doubted whether there was matter or not, almost conceiving that it preceded suppuration. It certainly has this effect long before there is any distention : besides, this of a pointing taking place, there is another effect of deep suppurations in consequence of inflammation, which is an œdematous appearance, or thickening of the superficial parts. This was taken notice of by Le Dran, in internal abscess of the abdomen, where adhesions had taken place, between the suppurating part and parietes of the abdomen, and by Mr. Pott, in suppuration of the brain ; whether in such there is a pointing I do not know.

As an abscess of the healthy kind requires but little surgical treatment, between its commencement and opening, it also requires but very little attention afterwards for the cure, or the restoring the parts.

It depends on the operation of the powers, or abilities the machine is in possession of more than any assistance the surgeon can give ; however, abscesses may have other circumstances attending them, besides soundness and unsoundness, which will require surgical treatment ; such as the extraction of exfoliated bones, which by their stay retard the cure. Farther, as few inflammations arise in perfectly sound parts and constitutions, it will generally be necessary to treat them in some degree as if they had an unsound tendency, and also according to other circumstances ; as no abscess can set about a cure till the matter is discharged, the first process, therefore, is the discharge of the matter ; but simply discharge is not always sufficient ; therefore it becomes necessary to consider whether or not, almost in every case it would not be proper to do more ; and I am inclined to believe that whatever would in general assist an unsound abscess, would also do the same to a sound one ; but this practice should be followed with great caution, and not carried too far ; for in many it will be perfectly unnecessary, therefore it should not be practised ; in others it will only be necessary in part ; besides, in many cases it may do harm, for many abscesses may have tolerable dispositions under the present treatment, yet may be in such a state as very readily to fall into an unsound one, of some kind or other, when too much violence is committed ; some having a tendency to irritability. On the other hand, our practice may fall short of the intention, as many parts have a strong tendency to indolence ; and if the stimulating method is applied to the first it would be unlucky, and vice versâ.

It will be generally more in the power of the parts to perform a cure if certain operations are done, which even dispose the most active and healthy disposition, both of constitution and of parts to heal sooner ; but this does not hold of the irritable. The first of these operations will be the mode of exposing abscesses, by opening them sufficiently, which will make any particular treatment afterwards, either less necessary, or more easy of application, if necessary ; so that the first principle of the cure, even of sound abscesses, may

be the freedom of opening them in the beginning; however, the more found they are, there is the lefs neceffity for
fuch treatment; for if it docs not give new powers to the
parts, it keeps up thofe of which they are already in poffeffion, and obliges them to go on towards a cure; for the
living principle in parts feems uneafy under the circumftance of expofure, and of having no fkin, more efpecially
found parts, therefore is roufed to action, acting with a
view to cover the part. It has no alternative; and as I
have juft now obferved that few fpontaneous abfceffes take
place from fo flight a caufe as fimple violence produces, there
muft be a fomething to be got the better of. This is perhaps, as well illuftrated in the fiftula in ano, as in any o-
ther; for without dividing along the gut to the bottom,
which is where the difeafe is, and where the abfcefs formed,
it feldom or never heals; however, all this will be according to circumftances, for if the fuppuration is quick and.
comes faft to the fkin, the parts will heal in the fame proportion more readily, either with or without opening; therefore, in fuch inftances, it is not fo neceffary to open freely,
though as it is not the method nature commonly takes, it
has by many been objected to; but let us obferve, that where
an abfcefs opens of itfelf by a fmall orifice, the parts are
commonly very found where the opening is, although the
bottom may be difeafed; but if it be difeafed where it opens,
then ulceration commonly takes place at this orifice, which
effect what fhould be done by art. To illuftrate that a large
opening is not detrimental to the healing of a fore, let us
obferve that there is no difference between an abfcefs opened largely, and a wound in confequence of an operation
which is not healed by the firft intention, fuch as an amputation, etc. for in fuch cafes there is a breach in the continuity of the parts communicating with the fkin, as large, if
not larger, than at the bottom, and it heals readily; we
endeavour, however, to remedy this as much as poffible by
faving fkin, which, in fome degree, anfwers to a fmall opening; and we may alfo obferve that where there is only a
fmall opening leading to a large cavity, which is to fuppurate, as in the cafe of an hydrocele treated by a cauftic or
featon, (which, when come to fuppuration, is in all refpects
fimilar to an abfcefs) that the whole fo far as fuppuration
extends, heals equally well with thofe that are wholly expofed; but I do not know that they do better; and where

the ſack is not very found, I do believe they do not do ſo well, as when more fully enlarged ; and we may alſo obſerve, that opening largely in the ſcrotum is not ſubject to the ſame inconvenience as in many other parts, for here there is ſo much looſe ſkin as to remove any retardment to the healing that might ariſe in other parts from opening largely ; however, after viewing this in every light, there ſeems but little advantage gained in the one way or the other. The opening more or leſs freely muſt be directed by ſome other circumſtance, by which the ſurgeon muſt be guided.

But as moſt abſceſſes owe ſome of their ſize to diſtention, and as this will be more or leſs according to circumſtances, it becomes neceſſary to diſtinguiſh the one kind from the other, for the one will require a freer opening than the other.

Abſceſſes in ſoft parts will owe more of their ſize to diſtention than thoſe in hard parts, ſuch as bones, joints, etc.

Abſceſſes in ſoft parts, not connected with the hard, will owe more of their ſize to diſtention than thoſe in ſoft parts connected with the hard ; for inſtance, an abſceſs in the calf of the leg, thick of the thigh, buttock, etc. will owe more of its ſize to diſtention than an abſceſs on the ſhinbone, on the head, etc. Therefore an abſceſs, whoſe ſize is in ſome degree owing to diſtention, need not be ſo freely opened as one that is not ; becauſe when the diſtention is taken off by the diſcharge of the pus, the parts will contract, or fall into their natural poſition, which cannot ſo eaſily happen in the other caſe. Beſides, the granulations will alſo be allowed to contract in the one much more than the other. However, we find many abſceſſes healing very readily without any other opening than what was at firſt made by ulceration, and this will be more readily effected if the abſceſs had been allowed to break of itſelf ; which I ſhall now more fully explain.

II.   OF THE TIME WHEN ABSCESSES SHOULD BE OPENED.

The natural procefs that abfceffes are obliged to go through for the difcharge of their contents is in general the moft proper, and it is fo much fo, as to be in moft cafes allowed to go on ; and this procefs becomes more neceffary in unfound abfceffes than in found ones, as it expofes them more fully, from ulceration having deftroyed more of the parts between the feat of the abfcefs and the external parts.

As abfceffes, wherever formed, muft increafe as they approach the fkin, and therefore increafe that part of their cavity next to the fkin, fafter than at the bottom, fo that they become in fome degree tapering towards the bottom, with a wide part immediately under the fkin, and this will be more or lefs fo, according to its depth, its meeting with different fubftances, which give a refiftance to the pus, or its coming faft or flowly to the fkin.

This fhape of the abfcefs, when allowed to take place is well adapted for healing, for it puts the bottom, which is the feat of the difeafe, more upon a footing with the mouth of the abfcefs, than it otherwife could be.   When thefe two are not well proportioned, there is a retardment in the cure ; for as the bottom, or part where the abfcefs begun, is more or lefs in a difeafed ftate, and as the parts between the feat of the abfcefs and the external furface are found parts, having only allowed a paffage for the pus, they of courfe have a ftronger difpofition to heal than the bottom has ; and we commonly find this to be the cafe.

If there could be made at any time a difference in the powers of healing between the mouth of the abfcefs and its bottom, it ought to be made the moft defective at the mouth of the abfcefs, as that part is the eafieft of management.   To have this effect produced as much as poffible, abfceffes fhould be allowed to go on till they break or open of themfelves ; for although abfceffes in general only open by a fmall orifice, more efpecially when found, yet it is to be remarked, that the fkin over the general cavity of the abfcefs is in fuch cafes fo much thinned as to have but very

little diſpoſition to heal, and is often ſo much ſo, as to ulcerate and make a free opening ; and if it does not, an o-
pening is more eaſily procured by art.

It is a curious circumſtance in the œconomy of abſceſs,
that thoſe that have the beſt diſpoſitions to heal come faſt-
eſt to the ſkin ; the lead takes place almoſt at a point, it
does not ſwell ſo much into that conical form, above-de-
ſcribed, not being under the ſame neceſſity in point of
healing, and it opens by a ſmall orifice ; while, on the o-
ther hand, if there is an indolence in the progreſs of the
abſceſs, it will ſpread more, or diſtend the ſurrounding
parts from their not being ſo firmly united by inflammation
in the one as they were in the other ; nor will ulceration
ſo readily take the lead, and it will come to the ſkin by a
broad ſurface, ſo as to. thin a large portion of the ſkin.
But abſceſſes ſhould only be allowed to open of themſelves
where the confinement of the matter can do no miſchief,
which will generally be in ſuch as ought to heal up from the
bottom ; but in the reduction of circumſcribed cavities to
the ſtate of an abſceſs, it will be in moſt caſes proper to
open early, as abſceſſes of the abdomen or thorax ; thoſe
within the cranium; thoſe of the eye ; and thoſe in joints.

In the abſceſs of the tunica vaginalis teſtis it would be
better to let it open of itſelf, as it ſhould be allowed to heal
up from the bottom, ſimilar to an abſceſs in the cellular
membrane.

If it ſhould be unneceſſary to open freely, or if from circumſtances this ſhould be impoſſible, it will in either caſe
be very proper to make the opening which is neceſſary or
practicable at the moſt depending part, with a view to remove
the preſſure ariſing from the matter collected, which is commonly called confinement or lodgement of matter, which will
otherwiſe happen ; for I ſhall obſerve, that a very ſmall
preſſure on that ſide of the abſceſs, next to the ſkin, may
produce ulceration there ; and although this preſſure in
many caſes might not be ſo great as to produce ulceration
at the bottom of the abſceſs, yet it may be ſo great as to
prevent granulations from forming on that ſide, and thereby retard the cure, as no union can take place but by means
of granulations ; or if it ſhould not prevent granulations
from forming, yet it might retard their growth, ſo that the
cure would be more tedious than if the preſſure did not

exift ; and this retardation will be greateft where the preffure is the greateft, which will be at the moft depending
part of the abfcefs ; fo that its upper part will readily heal
to a fmall point, and be reduced to the ftate of a fiftula.

But it is not always poffible to open at the moft depending part of an abfcefs, and when poffible, often very improper. When impoffible, perhaps, nothing more can be
done than to evacuate the matter as often as neceffary, and
by gentle preffure keep the fides of the finus together, to
allow their growing into one another ; but the fituation
will not in all cafes allow this.

The inexpediency of opening at the moft depending part
of an abfcefs will in general arife from the diftance between
the matter and the fkin at this part, for if the abfcefs is
pretty deeply feated, and points at a part fuperior to that of
its feat, which it fometimes does from the parts above,
being fuch as more eafily give way, in fuch a cafe it will
be proper to open it where it points ; for inftance, if an abfcefs is formed in the centre of the breaft, and opens at the
upper part, (which is often the cafe) it would be improper
to cut through the lower half, to allow the matter to
pafs that way, although it may make its way there afterwards, from the preffure of the matter, as was juft now
obferved ; which I have feen happen more than once.

If an abfcefs forms on the upper part of the foot, it is
improper to open through the fole of the foot to get at the
moft depending part of the abfcefs ; for befides cutting
fuch a depth of found parts, which is an objection, it would
be deftroying a great many ufeful parts. It would alfo be
impoffible to keep it open, the found parts having fuch a
difpofition to heal ; and it would be contradictory to my
firft pofition, which was to have parts as thin as poffible
before they are opened, in order to deftroy the healing difpofition there*.

As in fuch cafes, the place where the matter threatens
to open a paffage for itfelf, is where the future opening is
moft likely to be, and as the fituation is difadvantageous to
the healing of the feat of the abfcefs, it will be more neceffary to let it firft open of itfelf, becaufe the abfcefs juft un-

* One would imagine that this laft caution was hardly neceffary ; but I once faw a cafe where it was advifed upon the general principles of opening in the moft depending part.

der the ſkin will be increaſed in width, as was obſerved, and then to dilate it as freely as may be thought neceſſary; for by allowing abſceſſes to open of themſelves, the opening has a leſs diſpoſition to heal than if it had been opened early by art, therefore is more deſirable in ſuch ſituations.

III. OF THE METHODS OF OPENING ABSCESSES, AND TREATING THEM AFTERWARDS.

ALL abſceſſes, I have already obſerved, will open of themſelves, excepting where the matter is re-abſorbed; and I have alſo obſerved, that in general they ought to be allowed to open of themſelves, excepting ſome particular circumſtance calls for an early opening; but when the ſkin over the abſceſs is very thin, it is not of ſo much conſequence whether it is allowed to open of itſelf, or is opened at firſt by art.

In large abſceſſes it will generally be neceſſary to open them by art, whether they have opened of themſelves or not; for the natural opening will ſeldom be ſufficient for the complete cure; and although it may be ſufficient for the free diſcharge of the matter, yet they will heal much more readily if ſufficiently opened; for the ſkin over the cavity granulates but indifferently, and therefore unites but ſlowly with the parts underneath. Where the ſkin is very thin, looſe, and much of it, it may be neceſſary to remove an oval piece from the centre, where it is generally thinneſt. A queſtion naturally occurs, in what way ſhould theſe be opened?

The methods recommended and uſed are by inciſion and cauſtic. Inciſion may or may not remove a piece of the ſkin, but the cauſtic always will. I believe, as a general practice, there is no preference to be given to either;

but under circumſtances, the inciſion is beſt; for inſtance, where there is but little skin to ſpare, as on the
ſhin, ſcalp, etc. but where there is ſkin to ſpare, either ariſing from ſituation, as in the ſcrotum, or where a great
deal of skin was thinned, as in a great extent of inflamma-
mation and ſuppuration under the skin, a cauſtic will anſwer equally well; therefore I ſhould be very apt to be directed by my patients, if they had any fears or opinions
about the matter; for ſome have a terror at the idea of a
cutting inſtrument, while others hate the idea of a continued pain. If a cauſtic is approved of, then I ſhould prefer
the lapis infernalis, or ſepticus, to the common cauſtic;
the method of application I deſcribed, when ſpeaking of
the methods of producing death by art : but if left entirely
to myſelf, I ſhould prefer the inciſion to the cauſtic, becauſe it is immediately done.

If an abſceſs is allowed to open of itſelf, and this o-
pening is not enlarged, no dreſſing is neceſſary, nor any
thing but to keep the ſurrounding parts clean; the continuation of the poultice, which was before applied, (if convenient) is perhaps as good an application as any; and
when the tenderneſs ariſing from inflammation is over, then
lint and a pledget; but an abſceſs opened by a cutting inſtrument, may be called a mixed caſe, being both a wound
and a ſore, and is more of the nature of a freſh wound in
proportion to the thickneſs of the parts cut; and therefore the dreſſing ſhould be ſomewhat ſimilar to that of a
freſh wound. It is neceſſary that ſomething ſhould be put
into the opening, to keep it from healing by the firſt intention; if it is lint, it ſhould be dipt into ſome ſalve,
which will anſwer better than lint alone, as it will allow of
more early extraction; for ſuch ſores ſhould be dreſſed the
ſecond time the next day, or the ſecond day at lateſt; becauſe there is a ſuppurating ſore at the bottom, and the
pus requires being diſcharged much ſooner than if wholly
either a freſh wound or a circumſcribed cavity, which is to
ſuppurate, as the tunica vaginalis in the caſe of the radical
cure of the hydrocele. This pus keeps the lint (if dreſſed
with lint) moiſt, ſo that it does not dry, as in freſh wounds
in common. When the cut edges have come to ſuppuration, which will be in a few days, then the dreſſing afterwards may be as ſimple as poſſible, for nature will in general perform the cure.

If the abſcefs has been opened by cauſtic, and the ſlough is either cut out, or allowed to ſlough out, then it is to be conſidered as an entire ſuppurating ſore, and may be dreſſed accordingly ; perhaps dry liut is as good as any thing, till the nature of the ſore is known ; if of a good kind, the ſame dreſſing may be continued, but if not, then it muſt be dreſſed accordingly ; for nature cannot always peform a cure ; for parts which were at firſt found, or appeared ſo, from their readineſs to go through the firſt ſtages, will ſubſequently take on every ſpecies of diſeaſe, whether from indolence, from irritability, from ſcrofulous, or other diſpoſitions which in ſome caſes are produced from the nature of the parts diſeaſed, ſuch as bone, ligament, etc.

PART IV.

CHAPTER I.

OF GUN-SHOT WOUNDS.

━━━━━━━━━━

GUN-SHOT wounds may be faid to be an effect of a modern improvement in offence and defence, unknown in the former mode of war, which is ftill practifed where European improvements are not known ; and it is curious to obferve that fire-arms and fpirits are the firft of our refinement that are adopted in uncivilized countries ; and, indeed, for ages they have been the only objects that have been at all noticed or fought after by rude nations. It was not till the fourteenth century that gunpowder was made, or rather compounded ; but it was not till afterwards, applied to the purpofe of projecting bodies. But even now, the wounds received in war are not all gun-fhot wounds : fome, therefore, are fimilar in many refpects to thofe received in former times.

The knowledge of the effect of gunpowder, and its application to the art of war, or the projection of bodies for the deftruction of men, has been in fome degree accompanied by improvements in the arts and fciences in general, and among others, that of furgery, in which art, the healing of wounds fo produced, makes a material part. In France, more efpecially, the ftudy of both were carried to confiderable lengths ; but though the art of deftruction has been there improved and illuftrated by writings, it is rather furprifing that the art of healing fhould not have been equally illuftrated in the fame manner. Little has been written on this fubject, although, perhaps when we take every circumftance into confideration, it requires particu-

lar difcuffion ; and what has been written is fo fuperficial,
that it deferves but little attention. Practice, not precept,
feemed to be the guide of all who ftudied in this branch ;
and, if we obferve the practice hitherto purfued, we fhall
find it very confined, being hardly reduced to the common
rules of furgery, and therefore it was hardly neceffary for a
man to be a furgeon to practice in the army.

I. THE DIFFERENCE BETWEEN GUN-SHOT WOUNDS AND COMMON WOUNDS.

Gun-shot wounds are named, as is evident, from the
manner in which they are produced.   From the frequency
of their happening in the time of battle to a fet of men ap-
propriated for war, both by fea and land : and from the
appointment of particular furgeons for their cure, they have
been confidered apart from other wounds, and are now become almoft a diftinct branch of furgery.

Gun-fhot wounds are made by the projection of hard obtufe
bodies, the greateft number of which are mufket-balls; for cannon-balls, pieces of fhells and ftones from ramparts in fieges,
or fplinters of wood, etc. when on board of a fhip in an en-
gagement at fea, can hardly have their effects ranked among
gun-fhot wounds, they will come in more properly with
wounds in general.   As the wounds themfelves made by
thofe very different modes will in general differ very confiderably ; any peculiarity that may be neceffary in the treatment of gun-fhot wounds, from thofe made by cannon-balls,
fhells, etc. or even common wounds, will generally belong
to thofe made by mufket-balls.

The whole of gun-fhot wounds will come within the de-
finition of accidents.   They are a recent violence committed on the body ; but they often become the caufe of,
or degenerate into a thoufand complaints, which are the
objects of furgery or phyfic, many of which are common to
accidents in general, and to many other difeafes ; of this
kind are abfceffes, ulcerating bones, fiftulæ ; but fome are
peculiar to gun-fhot wounds, as calculi in the bladder from

the ball entering that viſcus, conſumpticn frcm wounds in
the lungs, which I believe rarely happens ; for I cannot
ſay I ever ſaw a caſe where ſuch an effect took place. But
it is the recent ſtate in which they are diſtinguiſhed, and in
which they are to be conſidered as a diſtinct object of treatment.

Wounds of this kind vary from one another, which will
happen according to circumſtances ; theſe variations will
be in general according to the kind of body projected, the
velocity of the body, with the nature and peculiarities of
the parts injured. The kind of body projected, I have obſerved, is principally muſket-balls, ſometimes cannon-balls,
ſometimes pieces of broken ſhells, and very often on board
of ſhip, ſplinters of wood. Indeed the effects of cannon-
balls on different parts of the ſhip, either the containing
parts, as the hull of the ſhip itſelf, or the contained, are
the principal cauſes of wounds in the ſailor ; for a cannon-
ball muſt go through the timbers of the ſhip before it can
do more execution than ſimply as a ball, (which makes it a
ſpent ball) and which ſplinters the inſide of the ſhip very
conſiderably, and moves other bodies in the ſhip, neither
of which it would do if moving with ſufficient velocity ;
muſket or cannon-balls ſeldom doing immediate injury to
thoſe of that profeſſion. The wounds produced by the three
laſt bodies will be more like many common and violent accidents, attended with much contuſion and laceration of
parts.

Gun-ſhot wounds, from whatever cauſe, whether from
a muſket-ball, cannon-ball, or ſhell, etc. are in general contuſed wounds, from which contuſion there is moſt commonly a part of the ſolids ſurrounding the wound deadened, as the projecting body forced its way through theſe ſolids, which is afterwards thrown off in form of a ſlough,
and which prevents ſuch wounds from healing by the firſt
intention, or by means of the adheſive inflammation, from
which circumſtance moſt of them muſt be allowed to ſuppurate. This does not always take place equally in every
gun-ſhot wound, nor in every part of the ſame wound ; and
the difference commonly ariſes from the variety in the velocity of the body projected ; for we find in many caſes, where
the ball has paſſed with little velocity, which is often the
caſe with balls, even at their entrance, but moſt common-

ly at the part laſt wounded by the ball, that the wounds
are often healed by the firſt intention.

Gun-fhot wounds, from the circumſtance of commonly
having a part killed, in general do not inflame ſo readily
as thoſe from other accidents ; this backwardneſs to inflame
will be in the proportion that the quantity of deadened parts
bear to the extent of the wound ; from which circumſtance
the inflammation is later in coming on, more eſpecially
when a ball paſſes through a fleſhy part with great velocity;
becauſe there will be a great deal deadened, in proportion
to the ſize of the wound ; therefore inflammation in gun-
fhot wounds is leſs than in wounds in general, where the
ſame quantity of miſchief has been done ; and this, alſo, is
in an inverſe proportion to the quantity of the parts deadened, as I have already explained in my introduction to inflammation, viz. that inflammation is leſs where parts are
to ſlough, than where parts have been deſtroyed by other
means.   On the other hand, where the ball has fractured
ſome bone, which fracture in the bone has done confiderable
miſchief to the ſoft parts, independent of the ball, then there
will be nearly as quick inflammation as in a compound
fracture of the ſame bone, becauſe the deadened part bears
no proportion to the laceration or wound in general.

From this circumſtance, of a part being often deadened,
a gun-fhot wound is often not completely underſtood at firſt;
for it is at firſt, in many caſes, impoſſible to know what
parts are killed, whether bone, tendon, or ſoft part, till the
deadened part has ſeparated, which often makes it a much
more complicated wound than at firſt was known or imagined; for it very often happens, that ſome viſcus, or a
part of ſome viſcus, or a part of a large artery, or even a
bone, has been killed by the blow, which does not fhow
itſelf till the ſlough comes away.   If, for inſtance, it is a
part of an inteſtine that has received a contuſion, ſo as to
kill it, and which is to ſlough, a new ſymptom will moſt
probably appear from the ſloughs being ſeparated, the contents of the inteſtine will moſt probably come through the
wound ; and probably the ſame thing will happen when
any other containing viſcus is in part deadened ; but thoſe
caſes will not be ſo dangerous, as if the ſame loſs had been
produced at firſt, for by this time all communication will
be cut off between the containing and contained parts ;
nor will it be ſo dangerous as when a confiderable blood-veſ-

ſel is deadened; for in this caſe, when the ſlough comes off'
the blood, getting a free paſſage into the wound, as alſo
out of it, probably death will immediately follow. If this
artery is internal, nothing can be done ; if in an extremity
the veſſel may be either taken up, or probably amputation
may be neceſſary to ſave the perſon's life; therefore an early attention ſhould be paid to accidents, where ſuch an e-
vent is poſſible. In caſe of a bone being deadened, an exfoliation takes place.

Gun-ſhot wounds are often ſuch as do much miſchief to
vital parts, the effects of which will be according to the nature of the parts wounded, and the violence of the wound ;
and alſo to parts, the ſoundneſs of which are eſſential either to the health of the whole, or to the uſes of the parts
wounded ; ſuch as ſome viſcus, whoſe contents are voided
through the opening, or joints, the diſpoſition of which is
ſlow to heal, and whoſe uſes are impeded when healed.

Gun-ſhot wounds often admit of being claſſed with the
ſmall and deep ſeated wounds, which are always of a particular kind reſpecting the cure.

The variety of circumſtances attending gun-ſhot wounds
is almoſt endleſs ; the following caſe may be given as an
example.

An officer in the navy was wounded by a piſtol-ball, in
the right ſide, about the laſt rib ; it entered about five inches from the navel, and appeared on the inſide of the ſkin
about two inches from the ſpinal proceſs, having paſſed, I
believe, in among the abdominal muſcles. The only remarkable thing that occured was, that the cellular membrane for ſome way about the paſſage of the ball was œdematous, and when I cut out the ball, air came out with it.

II. OF THE DIFFERENT EFFECTS ARISING FROM THE DIFFERENCE IN THE VELO-CITY OF THE BALL.

MANY of the varieties between one gun-ſhot wound and
another, ariſe from the difference in the velocity of the body projected ; and they are principally the following.

If the velocity of the ball is ſmall, then the miſchief is leſs in all of them; there is not ſo great a chance of their being compounded with fractures of the bones, etc. but if the velocity is ſufficient to break the bone it hits, the bone will be much more ſplintered than if the velocity had been very conſiderable ; for where the velocity is very great, the ball, as it were, takes a piece out; however, all this will alſo vary according to the hardneſs of the bone. In a hard bone the ſplinters will be the moſt frequent.

When the velocity is ſmall, the direction of the wound produced by the ball, will, in common, not be ſo ſtrait, therefore its direction not ſo readily aſcertained, ariſing from the eaſy turn of the ball.

When the velocity is ſmall, the deadened part or ſlough is always leſs ; for with a ſmall velocity, a ball would ſeem only to divide parts, while when the velocity is great, the contrary muſt happen ; from this circumſtance it is, that the ſlough is larger at that orifice where the ball enters than where it comes out ; and if the ball meets with a great deal of reſiſtance in its paſſage through, there will very probably be no ſlough at all at its exit, which will be therefore only a lacerated wound.

The greater the velocity of the ball, the cleaner it wounds the part, ſo much ſo as almoſt to be ſimilar to a cut with a ſharp inſtrument ; from which circumſtance it might be imagined, that there ſhould be a ſmaller ſlough ; but I ſuſpect, that a certain velocity given to the beſt cutting inſtrument, would produce a ſlough on the cut edges of the divided parts ; for the divided parts not giving way equally to the velocity of the dividing body, muſt of courſe be proportionally bruiſed.

Gun-ſhot wounds are attended with leſs bleeding than moſt others; however, ſome will be attended with this ſymptom more than others, even in the ſame part ; this ariſes from the manner in which the wound is produced : bleeding ariſes from a veſſel being cut or broken ; but the freedom of bleeding ariſes from the manner in which this is done if the artery is cut directly acroſs, and it is done by a ball paſſing with a conſiderable velocity, it will bleed pretty freely ; if bruiſed, and in ſome degree torn, then it will bleed leſs. When the velocity of the ball is ſmall, the veſſel will be principally torn, for, they will have time to ſtretc

before the continuity of their parts gives way ; but if it is
great, they will bleed more freely, because velocity will
make up for want of sharpness.

According to the velocity of the ball so is the direction.
When the velocity is great, the direction of the ball will be
in general more in a strait line than when it is small ; for
under such circumstances the ball more easily overcomes
obstructions; and therefore passes on in its first direction.

Velocity in the ball makes parts less capable of healing
than when it moves with a small velocity ; therefore gun-
shot wounds in pretty thick parts are in general later of healing at the orifice where the ball enters, than at the orifice
where it passes out ; because it becomes in some degree
a spent ball, the part having less slough, being only torn,
which will often admit of being healed by the first intention.

In cases where the ball passes through, and in such a direction as to have one orifice more depending than the o-
ther, I have always found that the depending orifice healed
soonest, and more certainly so if the ball came out that way,
and also if the ball had been pretty much spent in its paffage ; therefore it will require art to keep the depending
orifice open, if thought necessary ; but this circumstance of
its being a spent ball, will not always happen, because if
the person is near the gun when fired, the velocity of the
ball will be very little diminished in its progress through the
soft parts ; and therefore it will have nearly the same velocity on both sides.

This fact of the lower orifice healing soonest, is common
to all wounds, and I believe is owing to the tumefaction
which generally arises from the extravasated fluid always
descending to the lower part, and being retarded at the
lower orifice, it is as it were stopped there, and presses the
sides of the wound together, obliging it to heal, if the parts
have not been deadened ; this is evidently the case after
the introduction of the seaton in the hydrocele, especially
if the two orifices of the seaton are at some distance ; but
in the hydrocele, there is a more striking reason for it ; for
in this disease, the extravasated fluids are wholly detained
about the lower orifice, as there is no depending part for
the fluid to descend to.

III. OF THE DIFFERENT KINDS OF GUN-SHOT WOUNDS.

Gun-shot wounds may be divided into the fimple, and the compound.   Simple, when the ball paffes into, or through the foft parts only : the compound will be according to the other parts wounded.

The firft fpecies of compound, are thofe attended with fractures of the bones, or with the wound of fome large artery.

The fecond fpecies of compound wounds is, where the ball penetrates into fome of the larger circumfcribed cavities.   This laft, or penetrating wound, may be doubly complicated, or may be divided into two.   Firft, fimply penetrating ; and, fecondly, where fome vifcus or contained part, as the brain, lungs, heart, abdominal vifcera, etc. is injured ; all which cafes will be taken notice of in their proper places.

CHAPTER II.

OF THE TREATMENT OF GUN-SHOT WOUNDS.

——————————

IT has been hitherto recommended, and univerſally practiſed by almoſt every ſurgeon, to open immediately upon their being received, or as ſoon as poſſible, the external orifice of all gun-ſhot wounds made by muſket-balls; ſo much has this practice ben recommended, that they have made no diſcrimination between one gun-ſhot wound and another ; this would appear to have ariſen, and to be ſtill continued, from an opinion that gun-ſhot wounds have a ſomething peculiar to them, and of ·courſe are different from all other wounds, and that this peculiarity is removed by the oļening ; I own that I do not ſee any peculiarity. The moſt probable way of accounting for the firſt introduction of this practice, is from the wound in general being ſmall, and nearly of a ſize from one end to the other ; alſo the frequency of extraneous bodies being forced into theſe wounds by the ball, or the ball itſelf remaining there; for the way in which theſe wounds are made, is by the introduction of an extraneous body which is left there, if it has not made its way through, ſo that the immediate cauſe of the wound makes a lodgment for itſelf; often carrying before it cloathes, and even the parts of the body wounded, ſuch as the ſkin, etc. from hence it would naturally appear at firſt view, that there was an immediate neceſſity to ſearch after thoſe extraneous bodies, which very probably led the ſurgeon to do it ; and in general the impoſſibility of finding them, and even of extracting them when found, without dilatation, gave the firſt idea of o⁻ pening the mouths of the wounds ; but from experience, they altered this practice in part, and became not ſo deſirous of ſearching after theſe extraneous bodies ; for they found that it was oftner impoſſible to find them than could at firſt have been imagined, and when found that it was not poſſible to extract them, and that afterwards theſe bodies were brought to the ſkin by the parts themſelves, and thoſe that could not be brought to the external ſurface in this way, were ſuch as gave little or no trouble afterwards, ſuch as balls ; yet they altered this practice only ſo far as reſpected the attempt to extract extraneous bodies, for when they found from experience, that it was not neceſſary nor poſſible to extract theſe immediately, yet they did not ſee that it therefore was not neceſſary to take the previous or leading ſteps towards it.

The circumſtance I have mentioned, of gun-ſhot wounds being contuſed, obliges moſt of them to ſuppurate, becauſe in ſuch caſes there is more or leſs of a ſlough to be thrown off, eſpecially at the orifice made by the entrance of the ball ; there is, therefore, a freer paſſage for the matter, or any other extraneous ſubſtance, than the ſame ſized wound would have, if made by a clean cutting inſtrument, even if not allowed to heal by the firſt intention.

From all which, if there is no peculiarity in a gun-ſhot wound, I think this of dilating them as a general practice ſhould be rejected at once, even were it only for this reaſon, that few gun-ſhot wounds are alike, and therefore the ſame practice cannot apply to all.

This treatment of gun-ſhot wounds is diametrically oppoſite to a principle which is generally adopted in other caſes, although not underſtood as a general rule, which is, that very few wounds of any kind require ſurgical treatment at their commencement, excepting with an oppoſite view from the above, viz. to heal them by the firſt intention.

It is contrary to all the rules of ſurgery founded on our knowledge of the animal œconomy to enlarge wounds ſimply as wounds ; no wound, let it be ever ſo ſmall, ſhould be made larger, excepting when preparatory to ſomething elſe, which will imply a complicated wound, and which is to be treated accordingly ; it ſhould not be opened becauſe it is a wound, but becauſe there is ſomething neceſſary to be done, which cannot be executed unlefs the wound is enlarged.

This is common furgery, and ought alfo to be military furgery refpecting gun-fhot wounds.

As a proof of the inutility of opening gun-fhot wounds as a general practice, I fhall mention the cafes of four Frenchmen, and a Britifh foldier, wounded on the day of the landing of our army on the ifland of Bellifle ; and as this neglect rather arofe from accident than defign, there is no merit claimed from the mode of treatment.

Cafe I. A. B. was wounded in the thigh by two balls, one went quite through, the other lodged fomewhere in the thigh, and was not found while he was under our care.

II. B. C. was fhot through the cheft ; he fpit blood for fome little time.

III. C. D. was fhot through the joint of the knee : the ball entered at the outer edge of the patella, crofïed the joint under that bone, and came out through the inner condyle of the os femoris.

IV. D. E. was fhot in the arm : the ball entered at the infide of the infertion of the deltoid mufcle, paffed towards the head of the os humeris, then between the fcapula and ribs, and lodged between the bafis of the fcapula and fpinal proceffes, and was afterwards extracted. The man's arm was extented horizontally when the ball entered, which accounts for this direction.

Thefe four men had not any thing done to their wounds for four days after receiving them, as they had hid themfelves in a farm-houfe all that time after we had taken poffeffion of the ifland ; and when they were brought to the hofpital, their wounds were only dreffed fuperficially, and they all got well,

A grenadier of the 30th regiment was fhot through the arm, the ball feemed to pafs between the biceps mufcle and the bone ; he was taken prifoner by the French. The arm fwelled confiderably, they fomented it freely, and a fuperficial dreffing only was applied. About a fortnight after the accident he made his efcape, and came to our hofpital ; but by that time the fwelling had quite fubfided, and the wounds healed ; there only remained a ftiffnefs in the joint of the elbow, which went off by moving it.

I. OF THE PROPRIETY OF DILATING
GUN-SHOT WOUNDS.

It would be abfurd for any one to fuppofe that there is never occafion to dilate gun-fhot wounds at all; but it is certain there are very few in which it is neceffary. It will be impoffible to determine by any general defcription what thofe are that ought to be opened, and what thofe are that ought not, that muft be left in a great meafure to the difcretion of the furgeon, when once he is mafter of the arguments on both fides.

Some general rules may be given with regard to the more fimple cafes; but with regard to the more complicated, the particular circumftances of each cafe are the only guide; and they muft be treated according to the general principles of furgery.

Let us firft give an idea of the wound that would appear to receive no benefit from being dilated; and firft of the moft fimple wounds.

If a ball paffes through a flefhy part where it can hurt no bone in its way, fuch as the thick of the thigh, I own, in fuch a fimple wound, I fee no reafon for opening it; becaufe I fee no purpofe that can be anfwered by it, except the fhortening of the depth of the wound made by the ball, which can be productiv⁻ of no benefit. If the ball does not pafs through, and is not to be found, opening can be of as little fervice.

If the opening in the fkin fhould be objected to, as being too fmall, and thereby forming an obftruction to the exit of the flough, etc. I think that in general it is not; for the fkin is kept open by its own elafticity, as we fee in all wounds; this elafticity, mufcles and many other parts have not; and in general the opening made by a ball is much larger than thofe made by pointed inftruments; for I have already obferved, that there is often a piece of the fkin carried in before the ball, efpecially if it paffed with confiderable velocity, befides the circular flough; fo that there is really in fuch cafes a greater lofs of fubftance; therefore, whatever matter or extraneous body there is, when it comes to the fkin, it will find a free paffage out. Nor does the wound in the fkin in general heal fooner than the bottom; and, indeed, in many cafes not fo foon, becaufe the fkin is generally the part that has fuffered moft.

' However, this is not an abſolute rule, for the ſkin ſometimes heals firſt ; but I have found this to be the caſe as often where openings had been made, as in thoſe where they had not ; and this will depend upon circumſtances or peculiarities ; ſuch as the bottom being at a conſiderable diſtance, with extraneous bodies, and having no diſpoſi-tion to heal, tending to a fiſtula ; and I have obſerved in thoſe caſes, that the wound or opening made by the ſur-geon generally ſkinned to a ſmall hole before the bottom of the wound was cloſed, which brings it to the ſtate it would have been in, if it had not been dilated at all, eſpecially if there are extraneous bodies ſtill remaining ; for an extra-neous body cauſes and keeps up the ſecretion of matter, or rather keeps up the diſeaſe at the bottom of the wound, by which means the healing diſpoſition of its mouth is in ſome degree deſtroyed.

Let me ſtate a caſe of this laſt deſcription. Suppoſe a wound made with a ball ; that wound (from circumſtances) is not to heal in ſix months, becauſe the extrane-ous bodies, etc. cannot be extracted, or worked out ſooner, or ſome other circumſtance prevents the cure in a ſhorter time ; open that wound as freely as may be thought neceſ-ſary, I will engage that it will be in a month's time in the ſame ſtate with a ſimilar wound that has not been opened, ſo that the whole advantage (if there is any) muſt be before it comes to this ſtate ; but it is very ſeldom that any thing of conſequence can be done in that time, becauſe the extra-neous bodies do not come out at firſt ſo readily as they do at laſt, for the inflammation and tumefaction, which ex-tends beyond that very opening, generally keeps them in ; and if the wound is opened on their account at firſt, it ought to be continued to the very laſt. Upon the ſame principle, opening on account of extraneous bodies at firſt cannot be of ſo much ſervice as opening ſome time after ; for the ſuppuration, with its leading cauſes, viz. inflam-mation and ſloughing, all along the paſſage of the ball, makes the paſſage itſelf much more determined and more eaſily followed; for the want of which, few extraneous bodies are ever extracted at the beginning, excepting what are ſuperficial, ſmall, and looſe.

If the extraneous bodies are broken bones, it seldom happens that they are entirely detached, and therefore must loosen before they can come away ; also the bones in many cases are rendered dead, either by the blow or by being exposed, which must exfoliate, and this requires some time ; for in gun-shot wounds, where bones are either bruised or broke, there is most commonly an exfoliation, because some part of the bone is deadened, similar to the slough in the soft parts.

A reason given for opening gun-shot wounds is, that it takes off the tention arising from the inflammation, and gives the part liberty ; this would be very good practice if tention or inflammation were not a consequence of wounds; or it would be very good practice, if they could prove that the effects from dilating a part that was already wounded were very different, if not quite the reverse of those of the first wound, but as this must always be considered as an extention of the first mischief, we must suppose it to produce an increase of the effects arising from that mischief; therefore this practice is contradictory to common sense and common observation.

They are principally the compound wounds that require surgical operations, and certain precautions are necessary with regard to them, which I shall here lay down.

As the dilatation of gun-shot wounds is a violence, it will be necessary to consider well what relief can be given to the parts or patient by such an operation ; and whether without it more mischief would ensue ; it should also be considered what is the proper time for dilating.

But it will be almost impossible to state what wound ought, and what ought not be opened ; this must always be determined by the surgeon, after he is acquainted with the true state of the case and the general principles ; but from what has been already said, we may in some measure judge what those wounds are that should be opened, in order to produce either immediate relief, or to assist in the cure : we must have some other views than those objected to, we must see plainly something to be done for the relief of the patient by this opening, which cannot be procured without it, and if not procured, that the part cannot heal, or that the patient most probably must loose his life.

The practice to be recommended here will be exactly fimilar to the common practice of furgery, without paying any attention to the caufe as a gun-fhot wound.

One of the principal points of practice will be to determine at what period of time the dilatation fhould be made.

Firft, if the wound fhould be a flight one, and fhould require opening, it will be better to do it at the beginning, before inflammation comes on ; for the inflammation, in confequence of both, will be flight ; but in flight cafes dilatation will never be neceffary, except to allow of the extraction of fome extraneous body that is near. But if the wound is a confiderable one, and it fhould appear upon confideration, that you cannot relieve immediately any particular part, or the conftitution, then you can gain nothing by opening immediately, but will only increafe the inflammation, and in fome cafes the inflammation arifing from the accident and opening together, may be too much for the patient ; under this laft circumftance, it would be more advifable to wait till the firft inflammation ceafes, by which means the patient will ftand a much better chance of a cure, if not of his life ; therefore it is much better to divide the inflammations ; however, it is poffible that the inflammation may arife from fome circumftance in the wound, which could be removed by opening it ; for inftance, a ball, or broken bone preffing upon fome part whofe actions are either effential to the life of the part or the whole, as fome large artery, nerve, or vital part ; in fuch the cafe will determine for itfelf.

On the other hand, it may, in many cafes be better to remove the whole by an operation, when in fuch parts as will admit of it, which will be taken notice of.

Secondly, if an artery is wounded, where the patient is likely to become either too weak, or to loofe his life from the lofs of blood ; then, certainly the veffel is to be tied, and moft probably this cannot be done without previoufly opening the external parts, and often freely.

Or thirdly, in a wound of the head, where there is reafon to fufpect a fracture of the fcull, it is neceffary to open the fcalp, as in any other common injury done to the head where there was reafon to fufpect a fracture, and when opened, if a fracture is found, it is to be treated as any other fractured fcull.

Fourthly, where there are fractured bones in any part of
the body that can be immediately extracted with
advantage, and which would do much mischief if left,
this becomes a compound fracture wherever it is, and it
makes no difference in the treatment, whether the wound
in the skin was made by a ball, or the bone itself, at least
where the compound fracture is allowed to suppurate ; for
there is often a possibility of treating a compound fracture
as a simple one which gun-shot fractures, if I may be
allowed the expression, seldom will allow of ; but where
the compound fracture must suppurate, there they are very similar. However, there have been instances where a
fracture of the thigh-bone made by a ball has healed in the
same way as a compound simple fracture.

Fifthly, where there is some extraneous body which can
with very little trouble be extracted, and where the mischief by delay will probably be greater than that arising
from the dilatation.

Sixthly, where some internal part is misplaced, which
can be replaced immediately in its former position, such as
in wounds in the belly, where some of the viscera are protruded, and it becomes necessary to perform the operation
of gastroraphia ; which is to be done in this case in the
same manner as if the accident arose from any other cause ;
but the treatment should be different ; for gun-shot cases
cannot heal by the first intention, on account of the slough
that is to ensue.

Or, seventhly, when some vital part is pressed, so that
its functions are lost or much impaired, such as will often
happen from fractures of the scull, fractures of the ribs,
sternum, etc. in short, when any thing can be done to the
part after the opening is made for the present relief of the
patient, or the future good arising from it.   If none of
these circumstances has happened, then I think we should
be very quiet.   Balls that enter any of the larger cavities,
such as the abdomen or thorax, need not have their wounds
dilated, except something else is necessary to be done to the
contained parts, for it is impossible to follow the ball; therefore they are commonly not opened, and yet we find them
do very well.

Balls that enter any part where they cannot be followed
such as into the bones of the face need not have the wound
in the skin in the least enlarged, as it can give no assistance

to the other part of the wound, which is a bony canal. The following caſes are ſtrongly in proof of this, being reſpectively inſtances of both modes of practice.

CASE FIRST.

I was ſent for to an officer who was wounded in the cheek by a ball, and who had all the ſymptoms of an injured brain; upon examining the parts, I found that the ball had paſſed directly backwards through the cheek-bone; therefore, from the ſymptoms and from the direction of the wound, I ſuſpected that the ball had gone through the baſis of the ſcull into the brain, or at leaſt had produced a depreſſion of the ſcull there : I enlarged the external wound, and with my fingers could feel the coronoid proceſs of the lower jaw; I found that the ball had not entered the ſcull, but had ſtruck againſt about the temporal proceſs of the ſphenoid bone, which it had broke, and afterwards paſſed down on the inſide of the lower jaw. With ſmall forceps I extracted all I could of the looſe pieces of bone; he ſoon recovered from his ſtupor, and alſo from his wound. The ball afterwards cauſed an inflammation at the angle of the lower jaw, and was extracted. The good which I propoſed by opening and ſearching for extraneous bodies and looſe pieces of bone was the relieving of the brain; but as the ball had not entered the ſcull, and as none of the bones had been driven into the brain, it is moſt probable that I did no good by my opening; but that I could not foreſee.

CASE SECOND.

An officer received a wound by a ball in the cheek, (which in this caſe was on the oppoſite ſide) the wound led backwards, as in the other; by putting my finger into the wound I felt the coronoid proceſs of the lower jaw, as in the former; but he had no ſymptoms of an injured brain; I therefore adviſed not to open it, becauſe the reaſon for opening in the preceding caſe did not exiſt here; my advice was complied with, and the wound did well, and rather better than the former, by healing ſooner. The ball was never found, ſo far as I know.

The preſent practice is not to regard the balls themſelves, and ſeldom or ever to dilate upon their account, nor even to ſearch much after them when the wound is dilated, which shews that opening is not neceffary, or at leaft not made upon account of extraneous bodies.

This practice has arifen from experience; for it was found that balls, when obliged to be left, feldom or ever did any harm when at reft, and when not in a vital part; for balls have been known to lie in the body for years, and and are often never found at all, and yet the perfon has found no inconvenience.

This knowledge of the want of power in balls to promote inflammation when left in the body, arofe from the difficulty of finding them, or extracting them when found; and therefore in many cafes they were obliged to leave them.

One reafon for not readily finding the ball at firft is becaufe the parts are only torn and divided, without any lofs of fubftance, (till the flough comes off) by which means the parts colapfe and fall into their places again, which makes it difficult to pafs any thing in the direction of the ball, or even to know its direction. The different courfes they take, by being turned afide by fome refifting body, and alfo to the difficulty; as will be explained.

But the courfe of a ball, if not perpendicular, but paffing obliquely and not very deep a little way under the fkin, probably an inch or more, is eafy to be traced through its whole courfe, for the fkin over the whole paffage of the ball generally is marked by a reddifh line. I have feen this rednefs, even when the ball has gone pretty deep; it has none of the appearances of inflammation, nor of extravafation, for extravafation is of a darker colour, and what it is owing to, I have not been able to difcover. I can conceive it to be fomething fimilar to a blufh; only the fmall veffels allowing the red particles of the blood to flow more eafily.

II. OF THE STRANGE COURSE OF SOME BALLS.

THE difficulty of finding balls, I have juft obferved, often arifes from the irregular courfe they take. The regularity of the paffage of a ball will in general be in proportion to its velocity, and want of refiftance: for balls are turned afide in an inverfe proportion to the force that they come with; and this is the reafon why we feldom find them take a ſtrait courſe; for if they are ſpent balls, the ſoft parts alone is capable of turning them; and if they come with a conſiderable velocity, it is a chance they may hit ſome bone obliquely, and then they are alſo turned aſide, for any body that gives a ball the leaſt oblique reſiſtance, throws it out of its direct courſe; therefore, the balls that do not paſs through and through (which are the only ones that are ſearched after) will be in general ſpent ones, excepting thoſe that come directly againſt ſome conſiderable bone, as the thigh-bone, etc. As a proof that balls are eaſily thrown off obliquely, we often find that a ball ſhall enter the ſkin of the breaſt obliquely, and afterwards ſhall paſs almoſt round the whole body under the ſkin. The ſkin here is ſtrong enough to ſtop the balls coming out again, ſo that it turns it inwards, which meeting with the ribs, it is again turned out againſt the ſkin, and ſo on, alternately, as long as it has force to go on; however, in many caſes, the ball goes a little way after it has paſſed through the ſkin, and when it meets with any hard body on that ſide next the centre of the body, ſuch as a rib, its courſe is directed outwards, and it pearces the ſkin a ſecond time; but the velocity of ſuch balls muſt have been conſiderable.

I have ſeen a ball paſs in at one ſide of the ſhin-bone, and run acroſs it under the ſkin, without either cutting the ſkin acroſs, or hurting the bone; which ſhows that the velocity could not be great; for we know that there is not ſufficient room between theſe two parts in a natural ſtate for a ball to paſs; but the ball, after it had got under the ſkin, where there was room for it to cover itſelf, then came againſt the tibia, which threw it outwards, and the ſkin counteracting, it only raiſed the ſkin from the tibia, and paſſed on between them; but if this ball had had a ſufficient velocity, it would have either cut the ſkin acroſs, or taken a piece out of the bone, or moſt probably both.

Another circumſtance in favour of the uncertainty of their direction is, that the parts wounded are often not in the ſame poſition that they were when they received the ball. The French ſoldier who was wounded in the arm, was a ſtriking inſtance of this. The ball entered the arm about its middle, on the inſide of the biceps muſcle, and it was extracted from between the two ſcapulæ, cloſe on one ſide of the ſpinal proceſs of the back-bone. The reaſon of this ſtrange courſe, I have already obſerved in the caſe, was owing to

his having had his arm stretched out horizontally at the
time he was wounded, and the ball passed on in a stair line.

These uncertainties in the direction of the balls abovementioned, have made the common bullet-forceps almost
useless; yet forceps are not to be entirely thrown aside, for
it will often happen, that a ball will be found to lie pretty
near the external wound, which, if the ball was removed,
would heal, probably, by the first intention; for in such
superficial wounds they must have passed with little velocity; or if there was a part killed, it would heal immediately; but if there is a flough, this is best done after all inflammation, and the separation of the flough is over, for
then the passage of the ball is better ascertained, in confequence of the furrounding adhesive inflammation; and,
moreover, the granulations are beginning to push the extraneous body towards the furface; but the operation of ulceration, which brings it to the skin, being often too flow,
the ball, etc. had better be extracted, and even the part
might be dilated. However, I would be very cautious how
far I carried this practice, and only do it when all circumstances favoured.

For the fame reasons probes are become of little ufe; indeed, I think that they should never be ufed but by way of
satisfaction, in knowing sometimes what mischief is done;
we can perhaps feel if a bone is touched, or if a ball is near,
etc. but when all this is known, it is an hundred to one if
we can vary our practice in consequence of it. If the
wound will admit of it, the finger is the best instrument.

In cafes where the ball passes a considerable way under the
skin, and near to it, I think it would be advifeable to make
an opening midway between the two orifices which were
made by the ball (especially when the orifices are at a very
great diftance) that fractured bones, or extraneous bodies
may now, or hereafter be better extracted; for if this is
not done, we have often an abfcefs forming between them;
which, indeed, anfwers the fame purpofe, and often better;
but sometimes it should not be delayed for fuch an event
to take place.

Where the ball has passed immediately under the skin,
as in the cafe where the ball passed between the skin and
tibia, it will be often proper to open the whole length of
the passage of the ball, the necessity of which I think arif-

es from the skin not so readily uniting with the parts underneath, as muscles do with one another.

Although we have given up in a great measure the practice of searching after the ball, broken bones, or any other extraneous bodies, yet it often happens that a ball shall pass on till it comes in contact with the skin of some other part, and where it can be readily felt ; the question is, should such a ball be cut out ? if the skin is bruised by the ball coming against it, so that we may imagine that this part will slough off, in that case, I see nothing to hinder opening it, because the part is dead ; therefore no more inflammation can arise from the opening than otherwise would take place upon allowing the slough to be thrown off ; while, on the other hand, I should also suppose as little good to arise from it, because the ball will come out of itself when the part sloughs off; however, it may be suspected that before the slough falls off, the ball may so alter its situation, that it will be impossible to extract it by that opening ; however, I should very much suspect the ball altering its course under such circumstances, for if the skin was so much bruised as to slough, inflammation would soon come on, and confine the ball to that place ; however, it always gives comfort to the patient to have the ball extracted. But if the ball is only to be felt, and the skin quite sound, I would in that case advise letting it alone, till the wound made by the entrance of the ball had inflamed and was suppurating ; my reasons for it are these :

Firft, we find that moft wounds get well when the ball is left in (excepting it has done other mischief than simply paffing through the foft parts) and that very little inflammation attends the wound where the ball lodges, only that where it enters ; the inflammation not arifing fo much from the injury done by the ball, as from the parts being there expofed to the fuppurative inflammation, if it is immediately removed. There is always a greater chance of a slough where the ball enters than where it refts, arifing from the greater velocity of the ball ; for, beyond where the slough is, the parts unite by the firft intention.

Secondly, in thofe cafes where the ball paffes through and through, we have two inflammations, one at each orifice, inftead of the one at the entrance ; or a continued inflammation through and through, if the ball has paffed with great velocity. Where the ball makes its exit, the inflammation passes further along the passage of the ball, than when the wound has been healed up to the ball, and then cut out afterwards ; so that by opening immediately, the irritation will be extended further, and of course the disposition for healing will be prevented.

If this is the case, I think that two wounds should not be made at the same time ; and what convinces me more of it is, that I have seen cases where the balls were not found at first, nor even till after the patients had got well of their wounds ; and these balls were found very near the skin. They gave no trouble (or else they would have been found sooner) ; no inflammation came upon the parts, and they were afterwards extracted and did well.

Again, I have seen cases where the balls were found at first, and cut out immediately, which were similar to balls passing through and through ; the same inflammation came on the cut wounds that came on the wounds made by the entrance of the ball.

III. PENETRATING WOUNDS OF THE AB-
DOMEN.

Wounds leading into the different cavities of the body are very common in the army, and in a great measure peculiar to war ; they are mostly gun-shot wounds, but not always ; some being made with sharp-pointed weapons, as swords, bayonets, etc. they are pretty similar in whatever way they are made ; and I have given them a name expressive of the nature of the wound.   I shall not take notice of any of this kind, but those which penetrate into the larger cavities, as the abdomen, thorax, and scull ; but those into the scull are made most commonly by balls, shells, etc.

These wounds become more or less dangerous, according to the mischief done to the contents of the cavity into which they penetrate.

These wounds may be distinguished according as they are simply penetrating, without extending to the contained parts, or, as they affect these parts; and the event of these two kinds of wounds is very different; for in the first, little danger is to be expected if properly treated; but in the second, the success will be very uncertain; for very often nothing can be done for the patient under such wounds; and very often a good deal of art can be made use of with advantage.

Wounds of the parietes of the abdomen, not immediately inflicted on such a viscus as has the power of containing other matter, will in general do well, let the instrument that made the wound be what it will *. There will be a great difference, however, should that instrument be a ball passing with great velocity, for in this case a slough will be produced. But if it should pass with little velocity, then there will be less sloughing, and the parts will in some degree heal by the first intention, similar to those made by a cutting instrument; but although the ball has passed with such velocity as to produce a slough, yet that wound shall do well, for the adhesive inflammation will take place on the peritonæum all round the wound, which will exclude the general cavity from taking part in the inflammation, although the ball has not only penetrated, but has wounded parts which are not immediately essential to life, such as the epiploon, mesentery, etc. and perhaps gone quite through the body; yet it is to be observed, that wherever there is a wound, and whatever solid viscus may be penetrated, the surfaces in contact, surrounding every orifice, will unite by the adhesive inflammation, so as to exclude entirely the general cavity, by which means there is one continued canal wherever the ball or instrument has passed; or if any extraneous body should have been carried in, such as clothes, etc. they will also be included in these adhesions, and both these and the slough will be conducted to the external surface by either orifice.

---

* What I mean by a containing viscus, is a viscus that contains some foreign matter, as the stomach, bladder, ureters, gall-bladder, etc. to which I may add blood vessels.

All wounds that enter the belly, which have injured fome vifcus, are to be treated according to the nature of the wounded part, with its complications ; which will be many, becaufe the belly contains more parts of very diffimilar ufes than any other cavity in the body ; each of which will produce fymptoms peculiar to itfelf, and the nature of the wound.

The wounding of the feveral vifcera will often produce what may be called immediate and fecondary fymptoms, which will be peculiar to themfelves, befides what are common to fimple wounds, fuch as bleeding, which is immediate ; and inflammation and fuppuration, which are fecondary. Senfations alone will often lead to the vifcus wounded, and this is frequently one of the firft fymptoms.

The immediate fymptoms arifing from wounds in the different vifcera are as follows ;

From a wound in the liver there will be pain in the part, of the fickly or depreffing kind ; and if it is in the right lobe, there will be a delufive pain in the right fhoulder, or in the left fhoulder, from a wound in the left lobe.

A wound in the ftomach will produce great ficknefs and vomiting of blood, and fometimes a delirium ; a cafe of which I once faw in a foldier in Portugal, who was ftabbed into the ftomach with a ftiletto by a Portuguefe.

Blood in the ftools will arife from a wound in the inteftines, and according to the inteftine wounded, it will be more or lefs pure ; if the blood is from a high part of an inteftine it will be mixed with fœces, and of a dark colour ; if low, as the colon, the blood will be lefs mixed, and give the tinge of blood ; and the pain or fenfation will be more or lefs acute, according to the inteftine wounded : more of the fickly pain, the higher the inteftine, and more acute the lower.

There will be bloody urine from a wound in the kidnies or bladder ; and if made by fhot or ball, and a lodgment made, thefe bodies will fometimes become the caufe of a ftone. The fenfation will be trifling.

A wound of the fpleen will produce no fymptoms that I know of, excepting, probably, ficknefs, from its connexion with the nerves belonging to the ftomach, etc.

Extravafations of blood into the cavity of the abdomen will take place, more or lefs, in all penetrating wounds, and more especially if some viscus is wounded, as they are all extremely vascular ; and this will prove dangerous or not, according to the quantity.

These are the immediate and general symptoms upon such parts being wounded ; but other symptoms may arise in consequence of some of those viscera being wounded, which require particular attention. There may be wounds of the liver and spleen, which produce no symptoms but what are immediate, and may soon take on the healing disposition ; but wounds in those viscera which contain extraneous matter, such as the stomach, intestines, kidnies, ureters, and bladder, may produce secondary symptoms of a distinctive kind. If the injury is done by a ball to any of those viscera, the effect may be of two kinds ; one where it makes a wound, as stated above, the other where it only produces death in a part of any of them ; these will produce very different effects. The first will most probably be always dangerous ; the second will hardly ever be so. The first is, where the ball has wounded some one of the abovementioned viscera in such a manner as not to produce the symptoms already described, but produce one common to them all, viz. their contents or extraneous matter immediately escaping into the cavity of the abdomen. Such cases will seldom or ever do well, as their effect will hinder the abovementioned adhesions taking place. The consequence of this will be, that universal inflammation on the peritonæum will take place, attended with great pain, tension, and death. But all this will be in proportion to the quantity of wound in the part, and quantity of contents capable of escaping into the cavity of the abdomen ; for if the wound is small, and the bowels not full, then adhesions may take place all round the wound, which will confine the contained matter, and make it go on in its right channel. These adhesions may take place very early, as the following case shews.

The case of an officer who died of a wound which he received in a duel.

On Thursday morning, the 4th of September, 1783, about seven o'clock, an officer fought a duel in the Ring in Hyde-park, in which he exchanged three shot with his antagonist, whose last shot struck him on the right side, just below the last rib, and appeared under the skin on the oppofite fide, exactly in the correfponding place, and was immediately cut out by Mr. Grant.

About three hours after receiving this wound, I faw him with Mr. Grant. He was pretty quiet, not in much pain, rather low, pulfe not quick, nor full, and a fleepy languid-nefs in the eye, which made me fufpect fomething more than a common wound. He then had had neither a ftool nor made water, therefore it could not be faid what vifcera might be wounded. His belly had been fomented, a clyfter of warm water was ordered, and a draught with confec: card: as a cordial, with twenty drops of laudanum, to procure fleep, as he wifhed to have fome. We faw him again at three o'clock; the draught had come up. Had no ftool from the clyfter, nor any fleep; had made water, and no blood being found in it we conjectured that the kidnies, etc. were not hurt. He was now rather lower, pulfe fmaller, more reftlefs, a good deal of tenfion in the belly, which made him uneafy, and made him wifh to have a ftool. It was at firft imagined that this tenfion might be owing to extravafated blood; but on patting the belly, efpecially along the courfe of the tranfverfe arch of the colon, it plainly gave the found and vibration of air, there-fore we wifhed to procure a motion, to fee if we could not by that means have fome of that air expelled; we wifhed, alfo, to repeat the cordial and the opium, but the ftomach was become now too irritable to contain any thing, and was at times vomiting, independent of any thing he took; a clyfter was given, but nothing returned or came away. We faw him again at nine o'clock in the evening. His pulfe was now low and more frequent; coldnefs at times; vomiting very frequent, which appeared to be chiefly bile, with fmall bits of fomething that was of fome confiftence; the belly very tenfe, which made him extremely uneafy; no ftool. From nothing paffing downwards, and the colon continuing to fill, we began to fufpect that it was becoming paralitic, probably, from the ball having divided fome of its nerves.

Fumes of tobacco by clyfter were propofed, but we were loath to ufe it too haftily, as it would tend to increafe the difcafe, if it did not relieve; however, we were prepared for it.

Mr. Grant ftayed with him the whole night; all the a-bove fymptoms continued increafing, and about feven o'clock in the morning he died, viz. about twenty-four hours after receiving the wound.

He was opened next day at ten o'clock, twenty-feven hours after death, when we found the body confiderably putrid, although the weather was cold for the feafon, the blood having tranfuded all over the face, neck, fhoulders, and breaft, with a bloody fluid coming out of his mouth, with an offenfive fmell ; below this the body was not fo far gone.

On opening the abdomen, a good deal of putrid air rufhed out ; then we obferved a good deal of fluid blood, principally on each fide of the abdomen, with fome coagulum upon the inteftines ; when fponged up it might be about a quart.

The fmall inteftines were flightly inflamed in many places and there adhered. We immediately fearched for the paffage of the ball.

On fearching for the courfe of the ball, we found that it had paffed directly in, pierced the peritonæum, entered again the peritonæum, where it attaches the colon to the loins, paffed behind the afcending colon, and juft appeared at the right fide of the root of the mefentery where the colon is attached ; paffed behind the root of the mefentery, and entered the lower turn of the duodenum as it croffes the fpine ; then paffed out of that gut on the left of the mefentery, and in its courfe to the left fide, it went through the jejunum, about a foot from its beginning, then through between two folds of the lower part of the jejunum, taking a piece out of each, then paffed before the defcending part of the colon, and pierced the peritonæum of the left fide, as alfo fome of the mufcles, but not the fkin, and was immediately cut out, exactly in the fame place on the left, where it entered out the right ; fo it muft have paffed perfectly in an horizontal direction.

There was no appearance of extravafation of any of the contents of the inteftines loofe in the cavity of the abdomen. The inteftines in many places were adhering to one another, efpecially near to the wounds, which adhefions were recent, and of courfe very flight ; yet they fhewed a ready difpofition for union, to prevent the fecondary fymptoms, or what may be called the confequent, which would alfo prove fatal.

There was little or no fluid in the small intestines ; but there was a good deal of substance in consistence like fœces, in broken pieces, about the size of a nut through the whole track of the intestine, even in the stomach, which he vomited up ; but in the upper end of the jejunum, as also in the duodenum, there was some fluid mixed with the other ; but that fluid seemed to be rather bile. If this solid part was excrement then the valve of the colon must not have done its duty. Was all the thin part absorbed to hinder extravasation into the belly ? or was it all brought back into the stomach to be vomited up ? There was a good deal of air in the ascending, but more especially in the transverse turn of the colon.

This case admits of several observations and queries.

First, the lowness and gradual sinking, with the vomiting without blood, bespoke wounded intestines, and those pretty high up. It shews how ready nature is to secure all unnatural passages, according to the necessity.

Query, what could be the cause of his having no stool, even from the clyster ? Were the intestines inclinable to be quiet under such circumstances ? Would not he have lived if the immediate mischief had not been too much ? I think that if the immediate cause of death had not been so violent, nature would have secured the parts from the secondary, viz. the extravasation of the fœces.

What is the best practice where it is supposed an intestine may be wounded ? I should suppose the very best practice would be, to be quiet and do nothing, except bleeding, which in cases of wounded intestines is seldom necessary.

As he was extremely thirsty, and could not retain any thing in his stomach, which if he could, it would have been probably productive of mischief, by giving a greater chance of extravasation ; would not the tepid bath have been of service, to have allowed of fluids to enter the constitution ?

It is very possible that a wound of the gall-bladder, but more readily of the ductus communis, and also of the pancreatic duct, will produce the same effects, although not so quickly ; and it may be observed, that a wound in them could not be benefited by any adhesions that could take place, because the secreted fluids could never, most probably, get into the right channel again, and would therefore be the cause of keeping the external wound open, to difcharge the contents, as we find to be the cafe in the difeafe, called fiftula lacrimalis ; as alfo when the duct of the parotid gland is divided.

Of parts that have been only deadened.

Wounds will be very fimilar to the abovestated penetrating wounds, but they will differ from them in effects, arifing from a flough feparating from a containing vifcus ; for whenever the flough comes away, the extraneous or contained matter of that vifcus will efcape by the wound ; fuch as the contents of the ftomach, inteftines, ureters, bladder, etc. the two laft of which will be fimilar, or the flough may efcape by either of thefe outlets ; whereas, in the laft kind of wounds, any of the contents that could poffibly efcape would immediately pafs into the cavity of the abdomen.

The periods of thefe fymptoms appearing after the accident, will be according to the time of feparation ; which may be in eight, ten, twelve, or fourteen days.

This new fymptom, although in general very difagreeable, will not be dangerous*, for all the danger is over before it can appear ; but that the orifice fhould afterwards continue, and become either an artificial anus, or urethra, is a thing to be avoided ; though they commonly clofe up, and the contents are directed the right way ; in fuch cafe nothing is to be done, but dreffing the wounds fuperficially, and when the contents of the wounded vifcus become lefs, we may hope for a cure.

The following cafe explains the foregoing remarks.

A young gentleman was fhot through the body. The mufket was loaded with three balls, but there were only two orifices where they entered, and alfo only two where they came out, one of the balls having followed one of the others ; that there were three that went through him was evident, for they afterwards made three holes in the wainfcot behind him, but two very near one another.

The balls entered upon the left fide of the navel, one a little further out than the other, and they came out behind

---

* How far the contents of the ftomach efcaping through a wound might not be attended with bad confequences, I cannot pretend to fay.

pretty near the fpinal proceffes, about the fuperior vertebræ of the loins. From the clofenefs of the gun to the man when fired, which of courfe made it pafs with great veloci- ty, as alfo from the direction of the innermoft, which we fuppofed to be the double one, we were pretty certain that it had penetrated the cavity of the abdomen, but could not be fo certain of the courfe of the other.

The firft water he made after the accident was bloody, from which we knew the kidney was wounded ; but that fymptom foon left him.   He had no blood in his ftools, from which we concluded that none of the inteftines were wounded ; and no fymptoms of extravafation of the contents of any vifcus taking place, fuch as the fymptoms of in- flammation of the peritonæum, we were ftill more confirm- ed in our opinion.

The fymptomatic fever did not run higher than could have been expected ; nor was there more pain in the track of the ball than might be imagined.

Thefe confequent fymptoms of the immediate injury a- bated as foon as could be expected ; and in lefs than a fort- night, I pronounced him out of danger from the wound ; for no immediate fecondary fymptoms having taken place, I concluded that whatever cavities the balls had entered, there the furrounding parts had adhered, fo that the paf- fage of the ball was by this means become a complete canal; and therefore that neither any extraneous bodies that had been carried in with the balls, and had not been carried through, nor any flough that might feparate from the fides of the canal, nor the matter formed in it, could now get into the cavity of the abdomen, but muft be conducted to the external furface of the body, either through the wounds or from an abfcefs forming for itfelf, which would work its own exit fomewhere.

But this conclufion was fuppofed to be too hafty, and foon after a new fymptom arofe, which alarmed thofe who did not fee the propriety of my reafoning ; which was fome fæces coming through the wound ; this new fymptom did not alter my opinion, refpecting the whole operations of na- ture to fecure the cavity of the abdomen, but it confirmed it, (if a further confirmation had been neceffary) and there- fore I conceived it could not effect life; but as I faw the poffibility of this wound becoming an artificial anus, I was forry for it.   It was not difficult to account for the caufe of this new fymptom; it was plain that an inteftine, (the def-cending part of the colon moft probably) had only received a bruife from the ball, but fufficient to kill it at this part, and till the feparation of the flough had taken place, that both the inteftine and canal were ftill complete, and there-fore did not communicate with each other; but when the flough was thrown off, the two were laid into one at this part, therefore the contents of the inteftine got into the wound, and the matter from the wound might have got into the inteftine. However, this fymptom gradually decreaf-ed, by (we may fuppofe) the gradual contraction of this opening, and an entire ftop to the courfe of the fœces took place, and the wounds healed very kindly up.

But the inflammation, the fympathetic fever, the reducing treatment, and the fpare regimen, all tended to weaken him very much.

IV. OF PENETRATING WOUNDS IN THE CHEST.

Little notice has been taken of wounds in the cheft and lungs: indeed it would appear at firft, that little or nothing could be done; yet, in many cafes a great deal may be done for the good of the patient.

It is poffible a wound in the cheft may be of the firft kind, viz. only penetrating; yet from circumftances may prove fatal, as will be explained in the fecond or compli-cated, viz. a wound of the lungs.

It is pretty well known, that wounds of the lungs (ab-ftracted from other mifchief) are not mortal. I have feen feveral cafes where the patient has got well after being fhot quite through the body and lungs, while from a very fmall wound made by a fword or bayonet into the lungs, the pa-tients have died; from which I fhould readily fuppofe, that a wound in the lungs from a ball, would in general do better than a wound in the fame part with a pointed inftrument; and this difference in effects would appears in many cafes to arife from the difference in the quantity of blood extra-vafated; becaufe the bleeding from a ball is very inconfide-

rable in comparifon to that from a cut ; and there is therefore a lefs chance of extravafated blood, either in the cavity
of the thorax, or the cells of the lungs ; another circumftance that favours the gun-fhot wounds in thefe parts, is,
that they feldom heal up externally by the firft intention,
on account of the flough, efpecially at the wound made by
the entrance of the ball, fo that the external wound remains
open for a confiderable time, by which means any extravafated matter may efcape ; but even this has often its difadvan-
tages, for by keeping open the external wound, which leads
into the cavity, we give a chance to produce the fuppurative
inflammation through the whole furface of that cavity, which
moft probably would prove fatal, and which would be e-
qually fo if no vifcus was wounded ; but it would appear
that the cavity of the thorax does not fo readily fall
into this inflammation from a gun-fhot wound as we fhould
at firft imagine ; nor can we fuppofe that the adhefive inflammation readily takes place between the lungs and pleura
round the orifice, as we defcribed in the wounds of the abdomen, becaufe thefe parts are not under the fame circumftances that other contained, and containing parts are ; for
in every other cafe, the contained and containing have the
fame degree of flexibility, or proportion in fize. The brain
and the fcull have not the fame flexibility, but they bear
the fame proportion in fize. From this circumftance, the
lungs immediately collapfe, when either wounded themfelves or when a wound is made into the cheft, and not allowed heal by the firft intention, and become by much too
fmall for the cavity of the thorax, which fpace, muft be
filled either with air or blood, or both, therefore adhefion
cannot readily take place ; but it very often happens that
the lungs have previoufly adhered, which will frequently
be an advantage.

In the cafes of ftabs, efpecially if with a fharp inftrument,
the veffels will bleed freely, but the external wound will
collapfe, and cut off all external communication. If the
lungs are wounded in the fame manner, we muft expect a
confiderable bleeding from them, this bleeding will be in
the general cavity of the thorax (if the lungs at this part
have not previoufly adhered there) and likewife into the
cells of the lungs or bronchea, which will be known by producing a cough, and in confequence of it a bleeding at the
mouth ; for the blood that is extravafated into the air-cells

of the lungs, will be coughed up by the trachea, and by
that means will become a certain symptom of the lungs
being wounded ; but that which gets into the cavity of the
thorax cannot escape, and therefore must remain till the
absorbents take it up ; which they will do, if it is only in
small quantity ; but if in large quantity, this extravasated
blood will produce symptoms of another kind.

The symptoms of these accidents are,

First, a great lowness, which proceeds from the nature
of the parts wounded, and perhaps a fainting from the
quantity of blood lost to the circulation ; but this will be in
proportion to the quantity, and quickness with which it was
lost. A load in the breast will be felt, but more from a
sensation of this kind, than from any real weight ; and a
considerable difficulty in breathing.

This difficulty in breathing will arise from the pain the
patient will have in expanding the lungs in inspiration, and
will also proceed from the muscles of respiration of that
side being wounded, and this will continue for some time,
from the succeeding inflammation ; it will hinder the ex-
pansion of the thorax on that side, and of course in some
degree of the other side ; as we have not the power of
raising one side without raising the other * ; and if wounded by a cutting instrument, the lungs of that side not being
able to expand fully, by the cavity of the thorax being in
part filled with blood, will also give the symptoms of difficulty of breathing.

The patient will not be able to lie down, but must sit
upright, that the position may allow of the descent of the
diaphragm, to give room in the chest ; all which symptoms
were strongly marked in the following case.

A person received a stab behind the left breast with a
small sword ; the wound in the skin was very small. He
was almost immediately seized with a considerable discharge
of blood from the lungs, to near a quart, by the mouth,
which shewed that the lungs were considerably wounded,
for from the situation of the external wound we were sure
that the stomach could not be injured. His breathing soon
became difficult and painful, and his pulse quick ; he was

---

* I have often thought it a great pity, that we do not ac-
custom ourselves to move one side of our thorax independant of
the other, as we from habit move one eye-lid independant of
the other.

bled; thefe fymptoms increafed fo faft, that every one
thought him dying. He could only lie on his back, for if
he lay on the found fide, he could not breathe in the leaft,
and the pain would not let him lie on the unfound fide;
the eafieft pofition was an erect pofture, which obliged
him to fit in a chair for feveral days; when he coughed
he was in great pain, very feldom fpit with the cough, and
never difcharged any blood after the fecond day, by which
we fuppofed that the bleeding was ftopped in the lungs.

While the parts were in a ftate of inflammation he was
in great pain, his breathing exceffively quick, and his pulfe
hard and extremely quick; but as the inflammation went
off, he drew his breath in longer ftrokes, his pain became
lefs, and his pulfe not fo quick nor fo hard; but this laft
circumftance varied as he moved his body, coughed, or put
himfelf into a paffion, which he often did.

I fufpected from both the wound and its effects, that there
was a good deal of extravafated blood in the cavity of the
thorax; for I confidered that the blood which got out of
the veffels of the lungs into the wound in the lungs, would
find a readier paffage into the cavity of the thorax, than
into the cells of the lungs; and, indeed, every attempt to
the dilatation of the thorax would rather act as a fucker
upon the mouth of the wound in the lungs, as the preffure
of the external air was taken off by that means; I propofed
the operation for the empyema, becaufe the extravafated
blood muft comprefs the lungs of that fide, and hinder
their expanfion, and likewife irritate, and at laft might produce inflammation. He continued for fome days with little
variation, but upon the whole feemed getting better; but
the day before he died, he became worfe in his breathing,
which we imputed to his ftirring too much, and was rather
better on the day that he did die: juft before death he was
taken with a fort of fuffocation, and in half an hour he died.

Through his whole illnefs he had a moift fkin, and fometimes fweat profufely, at laft his legs fwelled.

At firft he only took a fpermaceti mixture with a little
opium, which gave him relief; I wanted to increafe the
opium, but it was objected to, for fear it fhould bind the
cheft too much, as it often does in afthmas, therefore it
was given with the fquills. On the day that he died, we
ordered him the bark with a fudorific.

As this was very different from a common afthma, and

the difficulty of breathing arising entirely from the inflammation of the intercoftal mufcles and lungs, and likewife from having but one lung, I thought it advifeable to give opium in this cafe, as it would take off the irritation of the inflamed parts, and therefore allow a greater ftretch or expanfion; efpecially as we found whenever it was given, that it gave relief, and produced thefe effects.

One might at firft wonder why he fhould breathe with fuch difficulty, as he had one fide whole or found; for I have feen people breathe pretty freely who have had but one fide to expand; but when we confider the cafe, we can eafily account for this.

After death we opened him; on raifing the fternum I cut into the cavity of the thorax, and a great deal of blood gufhed out at the incifion: we fponged out of the left fide of the thorax above three quarts of fluid blood; the coagulum appeared to have been attracted to the fides of the cavity every where as if it had been furred over with the coagulating lymph which was no where floating in the fluid; but moft probably the extravafated blood had.never coagulated, and this thick buff cruft was an exudation of coagulating lymph from the lungs and pleura which covers the ribs, as in all inflammations; if fo, this is another inftance, befides that of the inflammation of veins, in which the coagulating lymph coagulates immediately upon being thrown upon the furface, for if it had not, it then muft have mixed with the blood in the cheft, and only been found floating there.

The lungs were collapfed into a very fmall fubftance, and therefore firmer than common; we obferved the wound in them which corefponded to the wound in the pleura; I introduced a probe into the wound in the lungs, which paffed near four inches, but was not certain whether it did not make fome way for itfelf; however, I traced the wound by opening the lungs, and could eafily diftinguifh the wounded part by the coagulated blood that lay in it.

I found the heart and infide of the pericardium inflamed, and their furface furred over with coagulating lymph, fimilar to that on the lungs, The lungs of the right fide had alfo become a little inflamed on their anterior edges.

Wounds in the lungs generally become a caufe of a quick pulfe; this likewife may arife in fome degree from the lungs being fo immediately concerned in the circulation, that any thing that gives a check to the blood's free motion through them, may affect the heart. But the pulse becomes hard, which arises from the nature of the inflammation that attends, and also from the wound being in a vital part.

In the cases arising from balls, nothing in general is to be done but to keep quiet, and dress the wounds superficially; for any extravafated blood that might have got into the cavity of the thorax will generally make its escape by the external wound, as also any matter from suppuration. But in the cases of wounds made by cutting instruments, and where there is reason to suspect a considerable quantity of blood in the cavity of the thorax, then we may ask what should be done? and the natural answer is, that the operation for the empyema should be performed. This operation will relieve the patient and bring the disease to the simple wound, and somewhat nearer to the gun-shot wound; it should be performed as soon as possible, before the blood can have time to coagulate; for the coagulum of the blood may be with difficulty extracted.

The enlargement of the wound already made will often answer; but if that is in such a situation as to forbid dilatation, then the common directions for the empyema are to be followed here.

When all symptoms appear, and we have great reason to suppose a considerable extravafation of blood into the cavity of the chest, I think that we should not hesitate in performing the operation for the empyema.

V. OF CONCUSSIONS AND FRACTURES OF THE SCULL.

THESE injuries, in consequence of a musquet-ball, differ in nothing from the same accidents arising from any other cause, excepting the lodgement of the ball, which I imagine will require no peculiar mode of treatment.

VI. OF WOUNDS COMPOUNDED WITH FRAC-TURED BONES, OR CONTAINING EXTRA-NEOUS BODIES.

THE compound gun-shot wounds, where bones are broke, or where there are extraneous bodies that continue the irritation, similar to compound fractures, seldom or ever heal at once, or by regular degrees, as in the former, but generally heal very quick at first, upon the going off of the inflammation, similar to the healing of simple gun-shot wounds ; but when healed so far, as to be affected by the extraneous bodies, then they become slow in their progress, till at last they come to a stand, or become fistulous ; in which state they continue till the irritating cause is removed ; and this takes place even if the dilatation should have been made at first as large as could be thought necessary ; so that the opening at first, in such cases, can only let out those extraneous bodies or detached bones, that are perfectly loose, or becomes loose while the wound continues large; however, even this can only take place in superficial wounds; but in those that are deep, or where there is an exfoliation to take place, the dilated part always heals up long before they are fit to make their exit ; but before this happens the parts often acquire an indolent diseased state, and even when all extraneous bodies are extracted, the parts do not readily heal.

When a wound comes to this stage, surgeons generally put in sponge or other tents in the opening, or apply some corroding medicine to keep it open, and also with a view to make it wider ; but this practice is unnecessary, as a wound in such a state seldom heals entirely over, nor do tents add much to the width of the wound, and always confine the matter between the two dressings.

Where an exfoliation is expected, it is generally better to expose as much of the bone as possible ; it keeps up a kind of inflammation, which I imagine gives a disposition for this process. This can only be done where the bone is pretty superficial ; but in cases where the separation has already taken place, and it is now to make its way to the skin, like any other extraneous substance, then, in-

VOL. II.   N n

stead of the practice of sponge tents, to keep the orifice in
the skin open, it would be often better in such cases, to let
the whole heal over, because the extraneous body would
form an abscess round itself, which would enlarge the cavity, and produce the ulcerative inflammation quicker towards the surface ; and when that was opened, the extraneous body could be with more ease extracted, or would
come out of itself ; but this method of healing the mouths
of fistulous sores is not always practicable.

If this last practice has no inconveniences attending it, it
has this advantage, that the patient has not the disagreable
trouble of having a sore to dress every day, till the extraneous body comes away, which I think is no small consideration.   This practice, however, is not to be followed in
every case ; for instance, if the wound should communicate with a joint, as is common to most sores in the foot
and hand, where the bones are diseased, it would be, in
such cases very imprudent to allow the wound to heal, as
the confined matter would get more readily into the different joints, and increase the disease ; there may be other
causes to forbid this, as a general practice.

If wounds are to be kept open at their mouths, whose
bottoms have not a disposition to heal, they should be kept
open to that bottom ; because, whenever they do heal at
their mouths, it is most commonly owing to their sides underneath first uniting ; for the skin will seldom unite when
all beyond it is open.

In wounds that become fistulous, where there is no extraneous body, there is always a diseased bottom, which is
to be looked upon as having the same effect as an extraneous substance. To alter this diseased disposition, they should
be opened freely, as large openings produce quick inflammation, quick suppuration, and quick granulations,
which are generally found when they arise from such a
cause ; on the other hand, letting such wounds heal at their
mouths, has often a salutary effect, as it becomes a means
of destroying this diseased part by the formation of an
abscess there, and in general, there can be no better way
of coming at a part or extraneous body, than by the formation of an abscess there.   It is a natural way of opening
to relieve diseased parts ; but we often find in practice, that
this method is not sufficient, either for the extraction of
extraneous bodies, or to expose the diseased bottom, ex-

cepting thefe abfceffes are opencd very largely by art, fo as to expofe the whole of the difeafed parts or extraneous body.

VII. OF THE TIME PROPER FOR REMOV-
ING INCURABLE PARTS.

MANY gun-fhot wounds are at the very firft evidently in-curable, whether in a part that cannot be removed, or in one that will admit of being removed. When fuch wounds are in parts that will not admit of a removal of the parts in-jured, then nothing can be done by furgery ; but when in a part that can be removed, then a removal of the injured part is to be put in practice ; but even this is to be under certain reftrictions ; perhaps it fhould not be done immedi-ately upon the receiving of the injury, excepting where a confiderable blood-veffel is wounded, fo as to endanger the life of the perfon, and that it abfolutely cannot be taken up ; or it is fufpected that the inflammation, in confe-quence of the accident will kill ; by which means you have only the inflammation in confequence of the amputation ; but this is a bad recourfe, efpecially if it is a lower extre-mity this is to be amputated, and which is perhaps the on-ly part that can be removed of which the inflammation will kill.

How far the fame practice is to be followed in cafes which we may fuppofe will not kill ; but that the part is fo hurt, as to all appearance not to be in the power of furgery to fave, I will not now determine. This is a very different cafe from the former, and its confequences depend more upon contingences, fo that the part fhould be removed on-ly when the ftate of the patient in other refpects will ad-mit of it ; but this is feldom the cafe, for few people in full health are in that ftate, and ftill lefs fo thofe who are ufu-ally the fubjects of gun-fhot wounds ; the fituation they are in at the time, from the hurry of mind, makes it here in general to be the very worft practice ; it will in general therefore, be much better to wait till the inflammation, and all the effects of both the irritation and inflammation shall be gone off.

If these things are not sufficiently attended to, and the first inflammation, as in the first stated case, (for instance, that which is likely to prove mortal) is allowed to go on, the patient will most probably loose his life; or if the first inflammation is such as is likely to go off, according to the last stated case, then we should allow it to go off before we operate, and not run the risk of producing death by an operation; for I have already observed, few can support the consequences of the loss of a lower extremity when in full health and vigour: we know that a violent inflammation will in few hours alter the healthy disposition, and give a turn to the constitution, especially if a confiderable quantity of blood has been lost, which most probably will be the case where both accident and operation immediately succeed one another.

The patient under such circumstances becomes low, simply by the animal life loosing its powers, and hardly ever recovers afterwards.

After considering the curative treatment of gun-shot wounds, and other accidents common to the soldier, as also the sailor, let us further consider the treatment of those patients, whose wounds at the very first appear to be incurable, when they are in parts that will admit of being removed.

The operation itself is the same as in other cases, and the only subjects of peculiar consideration here are the situation of the patient, and the proper time for performing the operation after the injury.

I have already given some directions with regard to the proper time of operating, in treating upon the dilatation of gun-shot wounds, which are in some degree applicable here; but we shall consider this now more fully, as the proper time of removing a part is often much shorter than that of dilating.

Amputation of an extremity is almost the only operation that can, and is performed immediately on receiving the injury.

As these injuries in the soldier are generally received at a diftance from all care, excepting what may be called chirugical, it is proper we should consider how far the one should be practifed without the other. In general, furgeons have not endeavoured to delay it till the patient has been housed, and put in the way of a cure; and, there-fore, it has been a common practice to amputate on the field of battle: nothing can be more improper than this practice, for the following reasons. In such a situation it is almost impossible for a surgeon, in many instances, to make himself sufficiently master of the case, so as to perform so capital an operation with propriety; and it admits of dispute, whether at any time, and in any place, amputation should be performed before the first inflammation is over: when a case is so violent as not to admit of a cure in any situation, it is a chance if the patient will be able to bear the consequent inflammation, therefore, in such a case it might appear, at first sight, that the best practice would be to amputate at the very first; but if the patient is not able to support the inflammation arising from the accident, it is more than probable he would not be able to support the amputation and its consequences: on the other hand, if the case is such as will admit of being brought through the first inflammation, although not curable, we should certainly allow of it, for we may be assured, that the patient will be better able to bear the second.

If the chances are so even, where common circumstances in life favour the amputation, how must it be where they do not? how must it be with a man, whose mind is in the height of agitation, arising from fatigue, fear, distress, etc.? These circumstances must add greatly to the consequent mischief, and cast the balance much in favor of forbearance.

If it should be said, that agreeable to my argument, the same circumstances of agitation will render the accident itself more dangerous? I answer, that the amputation is a violence superadded to the injury; therefore, heightens the danger, and when the injury alone proves fatal it is by slower means.

In the first case, it is only inflammation; in the second, it is inflammation, loss of substance, and most probably loss of more blood, as it is to be supposed that a good deal has been lost from the accident, not to mention the awkward manner in which it must be done.

The only thing that can be said in favour of amputation on the field of battle is, that the patient may be moved with more ease without a limb, than with a shattered one; however, experience is the best guide; and I believe it is universally allowed by those whom we are to esteem the best judges, those who have had opportunity of making comparative obfervations, with men who have been wounded in the fame battle, fome where amputation had been performed immediately, and others where it had been left till all circumftances favoured the operation ; it has been found that few did well who had their limbs cut off on the field of battle ; while a much greater proportion have done well, in fimilar cafes, who were allowed to go on till the firft inflammation was over, and underwent amputation afterwards.

There will be exceptions to the above obfervations, which muft be in a great meafure left to the difcretion of the furgeon ; but a few of thefe objections may be mentioned, fo as to give a general idea of what is meant.

Firft, it is of lefs confequence, whichfoever way it is treated, if the part to be amputated is an upper extremity ; but it may be obferved, that there will be little occafion in general to amputate an upper extremity upon the field, becaufe there will be lefs danger in moving fuch a patient, than if the injury had happened to the lower.

Secondly, if the parts are very much torn, fo that the limb only hangs by a fmall connexion, then the circumftance of the lofs of fo much fubftance to the conftitution cannot be an objection, as it takes place from the accident; and, indeed, every thing that can poffibly attend an amputation; therefore, in many cafes, it may be more convenient to remove the whole. In many cafes it may be neceffary to perform the operation to get at blood-veffels, which may be bleeding too freely ; for the fearching after them may do more mifchief than the operation.

I have already obferved, that gun-fhot wounds do not bleed fo freely as thofe made by cutting inftruments, and are, therefore, attended with lefs danger of that kind ; however, it may often happen, that a confiderable veffel fhall be divided, and a confiderable bleeding take place ; in fuch cafes no time is to be loft, the veffels muft be taken up to prevent a greater evil : this operation may, in many cafes, be attended with confiderable trouble, efpecially as it will, in general, be on the field of action. Here the failor has the advantage of the foldier.

It will alfo be immediately neceffary on the field to replace many parts that would deftroy the patient if their reftoration was delayed, fuch as the bowels or lungs protruding out of their cavities; to remove large bodies, fuch as a piece of shell sticking in the flesh, which would give great pain, and do mischief by moving the whole together.

Very little can be done to relieve the brain in such a situation.

VIII. OF THE TREATMENT OF THE CONSTITUTION.

BLEEDING is recommended in gun-shot wounds, and in such a manner, as if of more service in them that wounds in general ; but I do not see this necessity more than in other wounds that have done the same mischief, and where the same inflammation, and other consequences are expected.

Bleeding is certainly to be used here, as in all wounds where there is a strong and full habit, and where we expect considerable inflammation and symptomatic fever ; but if it is such a gun-shot wound, as not to produce considerable effects, either local or constitutional, I would not bleed merely because it is a gun-shot wound ; and from what I have seen, I think that inflammation, etc. does not run so high in these wounds, as I should have at first expected. I believe this is the case with all contused wounds, where death in the part is a consequence : a contused wound is somewhat similar to the effects of a caustic ; for while the separation of the dead part is forming, the suppurative inflammation is retarded, and therefore not so violent ; but this can only be said of those wounds which are not complicated with any other injury, except what was produced by the balls passing through soft parts ; for if a bone is broke, it will inflame like any other compound fracture.

It is often of service in the time of inflammation to bleed in the part with leeches, or by punctures with a lancet ; this helps to empty the vessels of the part to lessen the in-

flammation fooner, and of courfe to promote fuppuration ;
but I muft own that bleeding muft be ufed with great
caution, where inflammation and fever run very high, for
to reduce the patient equal to the action at the time (which,
whether an increafed action, or an acquired one, is only
temporary) will be reducing him often too much for the
conftitution to fupport life, when this action ceafes ; for
the very worft thing that can happen, is the patient being
reduced too low ; we often afterwards find more difficulty
in keeping up with cordials, bark, etc. than we find in
lowering ; and we may avail ourfelves of obferving thofe
who have loft confiderable quantities of blood from the accident, which is always immediate, and we find too, that
a fecond bleeding, by fome other accident, although very
fmall in quantity, often deftroys our patient very quickly ;
but this will in a great meafure depend upon the feat of the
injury ; for in cafes of great violence done to fome parts of
our body, bleeding anfwers better than in others, becaufe
the fymptoms of diffolution, and diffolution itfelf, come on
fooner from mifchief done to fome parts, than when it is
done to others.

A man will bear bleeding better after an amputation of
the arm than leg ; better after a compound fracture of the
arm than the leg ; he will bear bleeding better after an injury done to the head, cheft, the lungs, etc. than to either
the arm or leg.

We find that injuries done to inactive parts, fuch as
joints, do worfe, and are more fufceptibie of irritation than
thofe in flefhy parts of the fame fituation.

It would appear upon the whole, that the decay of animal life is fooner brought on, when the inflammation is in
a part whofe circulation is not fo ftrong, and where the
nervous influence, or the force of the circulation is far
removed.

Bark is greatly recommended in gun-fhot wounds, and
with good reafon ; but it is ordered indifcriminately to all
patients that have received fuch wounds, whatever the
fymptoms or conftitution of the patient may be. That there
is no better medicine for wounds in general ; not only
when the inflammation is gone off, but in the time of inflammation, if the patient is rather low ; and, indeed, be-

fore it comes on, experience daily shews. Bark is to be looked upon as a strengthener, or regulator of the system, and an antispasmodic, both of which destroy irritation ; the bark and gentle bleedings, when the pulse begins to rise, are the best treatment that I know of in inflammations that arise either from accidents or operations ; one lessens the volume of the blood, and the increased animal powers at the time, which makes the circulation more free ; so that the heart labours less, and simple circulation goes on more freely ; the other gives to the blood that which makes it less irritating ; makes the blood-vessels do their proper offices, and gives to the nerves their proper sensations, which take off the fever.

Plate 1.

Fig. 1.

Fig. 2.

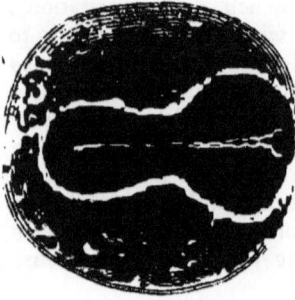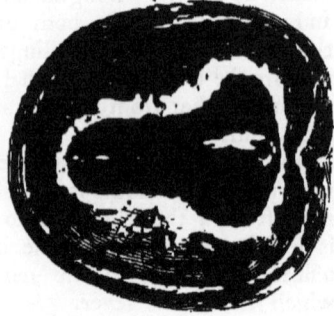

Fig. 3.

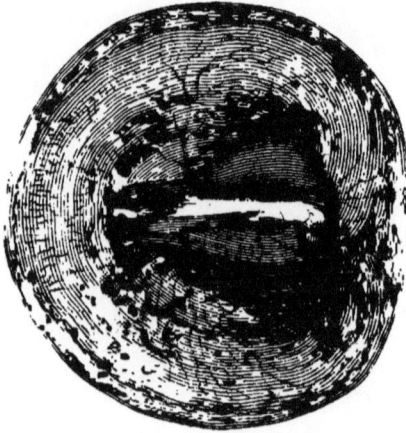

J. Trowhard

Plate II

Fig. 1.

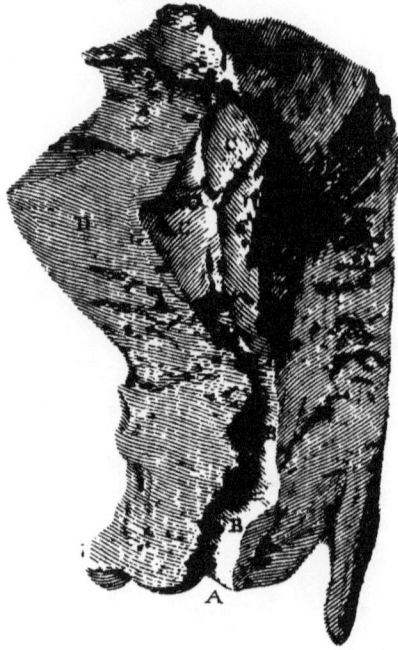

Fig. 2.

Vallang

Plate III

Fig. 1.

Fig. 2.

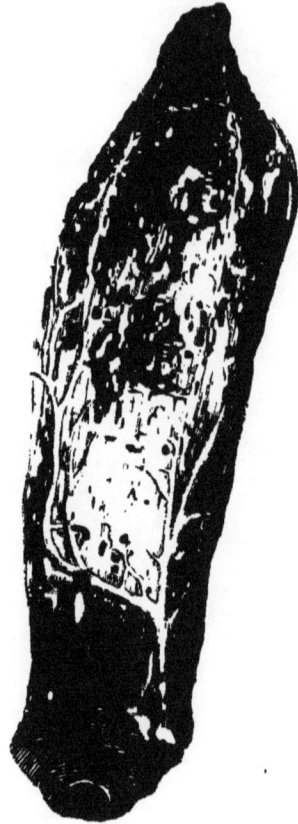

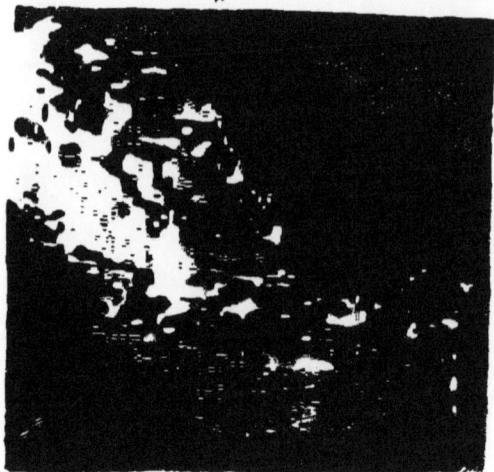

Fig. 1.

Fig. 2.

EXPLANATION OF THE PLATES.

PLATE FIRST.

In this plate is reprefented the embrio of the chick in the incubated egg, at three different ftages of its formation, beginning with the earlieft vifible appearance of diftinct organization. The preparations from which thefe figures are taken form part of a complete feries, contained in Mr. Hunter's collection of comparative anatomy. They are meant to illuftrate two pofitions laid down in this work, viz. That the bood is formed before the veffels, and when coagulated, the veffels appear to rife ; that when new veffels are produced in a part, they are not always elongations from the original ones, but veffels newly formed, which afterwards open a communication with the original.

FIGURE I.

In this figure the only parts that are diftinctly formed, are two blood-veffels ; on each fide of thefe is a row of fmall dots or fpecks of coagulated blood, which are afterwards to become blood veffels.

FIGURE II.

The formation of the embrio is further advanced, veffels appear to be rifing up fpontaneoufly in different parts of the membrane ; and the fpecks, out of which they are produced, are in many parts very evident.

FIGURE III.

The number of blood-veffels is very confiderably increaf-ed ; they now form a regular fyftem of veffels, compofed of larger trunks, and a vaft number of ramifications going off from them.

PLATE SECOND.

This plate reprefents a fection of the human uterus in the firft month after impregnation. The uterus itfelf is a little enlarged in fize, and thickened in its fubftance ; its cavity every where lined with a coagulum of blood, having a fmooth internal furface, but adhering firmly to the uterus.

The arteries are injected to fhew that it is uncommonly vafcular, and veffels are found to be injected in different parts of the coagulum.

The object of this plate is to fhew the readinefs with which veffels are formed in coagulated blood, when attached to a living furface, and its vafcularity being to anfwer ufeful purpofes in the machine ; of which this is a remarkable inftance, as it is to form the outer membrane of the fœtus or the connecting medium between it and the uterus.

FIGURE I.

A longitudinal fection of the uterus, in which the cavity is expofed.

A. The os tincæ projecting into the vagina, of which there is a fmall portion, to fhew the length to which the os tincæ projects.

BB. The cervix uteri.

CCC. The coagulated blood fmooth upon its internal furface, although extremely irregular.

DD. The cut furface of the fubftance of the uterus, which has fo intimate a connexion with the coagulum that the one appears to be continued into the other. The laminated appearance is produced by the fection of enlarged veins in a collapfed ftate, which are extremely numerous.

FIGURE II.

Is a thin flice of the fubftance of the uterus and the coagulum adhering to it, dried, and viewed in a microfcope, to fhew the vafcularity of the uterus, whofe veffels are diftinctly feen, continued into the coagulum, and paffing about half way through its fubftance.

PLATE THIRD.

REPRESENTS two rabbits' ears, one in the natural ftate, the other in an inflamed ftate, in confequence of having been frozen and thawed.

The veffels are injected, and as they belonged to the fame head, the force applied, and other circumftances muft have been exactly fimilar.

The difference in the fize of the veffels, and the difference in the thicknefs of the ears themfelves, is very evident ; but there was an opacity in the inflamed ear compared with the other, which it was not poffible to exprefs.

FIGURE. I.

The ear is in its natural ftate.
AA. The projecting part of the ear.
B. That part which is covered by the fkin of the head.
CCC. The principal arterial trunk.

FIGURE II.

The inflamed ear.
AA. B. CCC. reprefent the fame parts as in figure one.
D. A branch rather larger than the trunk, not diftinguifhable in the natural ftate of the ear.

PLATE FOURTH.

FIGURE I.

A PORTION of the illium taken from the inteftines of an afs. The inteftine was in a ftate of inflammation, and fhews the internal furface of the gut partly covered by a layer of coagulating lymph thrown out by the great degree of inflammation which the parts had undergone.

The internal membrane was extremely vafcular, and when injected, veffels were feen in portions of the coagulating lymph.

AA. The inner furface of the inteftine.
BB. The coagulating lymph which adhered to it.

FIGURE II.

The peritoneal coat of a portion of the human inteftine, in an inflamed ftate, to fhew its vafcularity, and to fhew a fmall portion of coagulating lymph attached to it by a narrow neck, which is fupplied with veffels from it.

F I N I S.